The Data-Driven Blockchain Ecosystem

This book focuses on futuristic approaches and designs for real-time systems and applications, as well as the fundamental concepts of including advanced techniques and tools in models of the data-driven blockchain ecosystem.

The Data-Driven Blockchain Ecosystem: Fundamentals, Applications, and Emerging Technologies discusses how to implement and manage processes for releasing and delivering blockchain applications. It presents the core of blockchain technology, IoT-based and AI-based blockchain systems, and various manufacturing areas related to Industry 4.0. The book illustrates how to apply design principles to develop and manage blockchain networks and also covers the role that cloud computing plays with blockchain applications.

All major technologies involved in blockchain-embedded applications are included in this book, which makes it useful to engineering students, researchers, academicians, and professionals interested in the core of blockchain technology.

The Data-Driven Blockchain Ecosystem
Fundamentals, Applications, and Emerging Technologies

Edited by
Alex Khang, Subrata Chowdhury,
and Seema Sharma

CRC Press is an imprint of the
Taylor & Francis Group, an **informa** business

MATLAB® is a trademark of The MathWorks, Inc. and is used with permission. The MathWorks does not warrant the accuracy of the text or exercises in this book. This book's use or discussion of MATLAB® software or related products does not constitute endorsement or sponsorship by The MathWorks of a particular pedagogical approach or particular use of the MATLAB® software.

First edition published 2023
by CRC Press
6000 Broken Sound Parkway NW, Suite 300, Boca Raton, FL 33487-2742

and by CRC Press
4 Park Square, Milton Park, Abingdon, Oxon, OX14 4RN

CRC Press is an imprint of Taylor & Francis Group, LLC

© 2023 selection and editorial matter, Alex Khang, Subrata Chowdhury, and Seema Sharma; individual chapters, the contributors

Reasonable efforts have been made to publish reliable data and information, but the author and publisher cannot assume responsibility for the validity of all materials or the consequences of their use. The authors and publishers have attempted to trace the copyright holders of all material reproduced in this publication and apologize to copyright holders if permission to publish in this form has not been obtained. If any copyright material has not been acknowledged please write and let us know so we may rectify in any future reprint.

Except as permitted under U.S. Copyright Law, no part of this book may be reprinted, reproduced, transmitted, or utilized in any form by any electronic, mechanical, or other means, now known or hereafter invented, including photocopying, microfilming, and recording, or in any information storage or retrieval system, without written permission from the publishers.

For permission to photocopy or use material electronically from this work, access www.copyright.com or contact the Copyright Clearance Center, Inc. (CCC), 222 Rosewood Drive, Danvers, MA 01923, 978-750-8400. For works that are not available on CCC please contact mpkbookspermissions@tandf.co.uk

Trademark notice: Product or corporate names may be trademarks or registered trademarks and are used only for identification and explanation without intent to infringe.

Library of Congress Cataloging-in-Publication Data
Names: Khang, Alex, editor. | Chowdhury, Subrata, editor. | Sharma, Seema, (Computer scientist), editor.
Title: The data-driven blockchain ecosystem : fundamentals, applications, and emerging technologies / edited by Alex Khang, Subrata Chowdhury, and Seema Sharma.
Description: First edition. | Boca Raton, FL : CRC Press, 2022. |
Includes bibliographical references and index.
Identifiers: LCCN 2022027699 (print) | LCCN 2022027700 (ebook) |
ISBN 9781032216249 (hbk) | ISBN 9781032216256 (pbk) | ISBN 9781003269281 (ebk)
Subjects: LCSH: Blockchains (Databases)
Classification: LCC QA76.9.B56 D38 2022 (print) |
LCC QA76.9.B56 (ebook) | DDC 005.74–dc23/eng/20220805
LC record available at https://lccn.loc.gov/2022027699
LC ebook record available at https://lccn.loc.gov/2022027700

ISBN: 978-1-032-21624-9 (hbk)
ISBN: 978-1-032-21625-6 (pbk)
ISBN: 978-1-003-26928-1 (ebk)

DOI: 10.1201/9781003269281

Typeset in Times
by Newgen Publishing UK

Contents

Preface ix
Acknowledgments xi
List of Editor Biographies xiii
List of Contributors xv

Chapter 1 Comprehensive Analysis of Fundamentals, Innovation, and
Key Challenges of Blockchain 1

*Mukesh Soni, Mohamed A. Elashiri, Abdulah S. Almahayreh,
and Abdelwahab Said Hassan*

Chapter 2 Cryptocurrency Methodologies and Techniques 21

S. Hasan Hussain, T. B. Sivakumar, and Alex Khang

Chapter 3 Framework for Modeling, Procuring, and Building Systems for
Smart City Scenarios Using Blockchain Technology and IoT 31

Rajendra Kumar, Ram Chandra Singh, and Rohit Khokher

Chapter 4 Development of a Framework Model Using Blockchain to
Secure Cryptocurrency Investment 51

*K. Mohamed Jasim, Manivannan Babu,
and Chinnadurai Kathiravan*

Chapter 5 A Blockchain Approach to Improving Digital Linked
Management Information Systems (MIS) 61

B. Akoramurthy, K. Dhivya, G. Vennira Selvi, and M. Prasad

Chapter 6 A Perspective on Blockchain-based Cryptocurrency to
Boost A Futuristic Digital Economy 83

S. Shivam Gupta, Sarishma Dangi, and Sachin Sharma

Chapter 7 Application of Blockchain in Online Learning: Findings in
Higher Education Certification 103

Pushan Kumar Datta and Susanta Mitra

Chapter 8	Robot Process Automation in Blockchain 113	
	R. K. Tailor, Ranu Pareek, and Alex Khang	

Chapter 9 A Novel Approach to Cryptography: Deep Learning-based Homomorphic Secure Searchable Encryption for Keyword Searches in the Blockchain Healthcare System 127

T. B. Sivakumar and Hasan Hussain

Chapter 10 Design and Implementation of a Smart Healthcare System Using Blockchain Technology with A Dragonfly Optimization-based Blowfish Encryption Algorithm 137

Shivlal Mewada, Dhruva Sreenivasa Chakravarthi, S. J. Sultanuddin, and Shashi Kant Gupta

Chapter 11 Implementation of a Blockchain-based Smart Shopping System for Automated Bill Generation Using Smart Carts with Cryptographic Algorithms 155

Parin Somani, Sunil Kumar Vohra, Subrata Chowdhury, and Shashi Kant Gupta

Chapter 12 Multi-Node Data Privacy Audit for Blockchain Integrity 169

A. Shenbaga Bharatha Priya, Sanjaya Kumar Sarangi, S. Balasubramanian, and Bhaskar Roy

Chapter 13 IoT, AI, and Blockchain: An Integrated System Investigation for Agriculture and Healthcare Units 189

Mandeep Singh, Ruhul Amin Choudhury, and Sweta Chander

Chapter 14 Security and Privacy Challenges in Blockchain Application 207

Khalid Albulayhi and Qasem Abu Al-Haija

Chapter 15 Blockchain-based Cloud Resource Allocation Mechanisms for Privacy Preservation 227

Akhilesh Kumar, Nihar Ranjan Nayak, Samrat Ray, and Ashish Kumar Tamrakar

Chapter 16 Blockchain-based Privacy Protection Credential Model for
Zero-Knowledge Proof over Distributed Systems............................247

*Sarfraz Fayaz Khan, Sumit Kumar, Ramya Govindaraj,
and Sagar Dhanraj Pande*

Index 267

Preface

The term *blockchain technology* is used in various ways. Some people use this term as bitcoin, some use it in cryptography, sometimes it is related to Ethereum or some related it to smart cards, etc. Sometimes we ought to consider blockchain as one more class of thing like the Internet – an exhaustive data innovation with layered specialized levels. But there is a confusion for many about the word "blockchain."

In simple langauge we can say that blockchain is a record-keeping technology which has been designed to protect the system from hacking, which is essential in today's scenario. It has been used in numerous applications like Education, Banking, Finance, Supply Chain, Healthcare, IoT, Transporation, Smart City, Agriculture, etc.

Nowadays, blockchain technology is growing faster day by day. Many websites explain blockchain technology but still there is a lack of comprehensive guidelines for terminology used in it. So in this context it has been decided that there should be a complete and comprehensive guide that covers all the terminology used in blockchain.

One of the major goals and objectives of this book is to give an extensive prologue to the hypothetical and reasonable parts of blockchain innovation. This book consists of all the materials essential for the complete understanding of blockchain innovation. The topics are organized in such a way that anyone can easily understand.

After reading this book, readers will actually be able to understand the inward activities of blockchain innovation and create blockchain applications. This book covers all topics pertinent to blockchain innovation, including Cryptography, Cryptocurrencies, Security, Ethereum, and the platforms and tools through which it can be perfectly utilized for blockchain advancement.

It is suggested that readers should have a basic idea of computer science and programming skills experience to benefit completely from this book. Even if that isn't the case, this book can still be effectively studied as significant foundation material is given where required.

Happy reading!

The Editors,
Alex Khang
Subrata Chowdhury
Seema Sharma

Acknowledgments

We are living in a world in which blockchain technology is crucial in every field and over the next few years the blockchain network will continue to change our lives; as it slowly rolls out, the IoT-based blockchain network will provide even better connectivity, and blockchain technology will indeed look pretty interesting.

Having an idea, turning it into a chapter, and then sharing it with us can be daunting for a contributor, but your effort and experience are both internally challenging and rewarding for the academic world. Without the support and contribution of friends and colleagues, this book would not exist. So, we especially want to say thanks to the individuals that have contributed to make this book successful.

To all the reviewers we have had the opportunity to collaborate with and appreciate their hard work from afar, we acknowledge their tremendous support and valuable comments. We would like to say a big thank you for being the inspiration and foundation for this project's success.

We thankfully acknowledge all the advice, support, motivation, sharing, and inspiration we have received from our faculty and academic colleagues.

We express our grateful thanks to our publisher CRC Press (Taylor & Francis Group) and the entire editorial team who have provided wonderful support in the timely processing of this manuscript and bringing out the book.

Sincerely, this is all about trust, honor, and respect for the contribution of all those who submitted the book chapters. We will always welcome the chance to represent you.

Thank you to all.

The Editorial Team

Editor Biographies

Alex Khang [PH] Doctor of Computer Science, Software Development Expert, AI and Data Scientist, Professor of IT in Universities of Science and Technology in Vietnam & United States, Chief of Technology (AI and Data Science Research Center) Global Research Institute of Technology and Engineering, NC, USA; 20+ Session Chair, 25+ International Keynote Speaker, 100+ Journal Paper Reviewer and Evaluator (Springer, CRC, Inderscience, IEEE, IGI Global, and Emerald). He has 25+ years of experience teaching Information Technology (Software Development, Database Technology, AI Engineering, Data Engineering, Data Science, Data Analytics, IoT, and Cloud Computing) in universities of technology in Vietnam, India, and the USA. He is the recipient of Best Professor of the Year 2021, Researcher of the Year 2021, Global Teacher Award 2021 (AKS), Life Time Achivement Award (Educacio World), and many other awards. He has contributed to various research activities in the fields of AI and Data Science while publishing articles in renowned journals and conference proceedings. He has published 52 IT books, as well as 6 AI and Data Science books. He has delivered many expert techtalks in Computer Science, both in academic and other professional contexts. He has over 28 years of experience in the field of software productions and has specialized in data engineering for foreign corporations from Germany, Sweden, the United States, Singapore, and multinational corporations. He is a specialist in data engineering and artificial intelligence at IT Corporation and also in the contribution stage of knowledge and experience into the scope of tech talk; he is a mentor, a consultant, a part-time lecturer, and an evaluator for 15+ Master's/Ph.D theses for local and international institutes and schools of technology.

Subrata Chowdhury is Assistant Professor & IQAC Coordinator, MCA, Sri Venkateswara Engineering College of Engineering & Technology (Autonomous), Andhra Pradesh, India. He works in the CSE Department of Sri Venkateshwara Engineering College. He has been working in the IT Industry for more than 5 years in R&D departments and has handled many projects in the industry with much dedication and keen timeliness. He has been handling projects related to AI, blockchains, and Cloud Computing for companies with various national and international clients. He had published 4 books from 2014 to 2019 in the domestic market and internationally for publishers CRC Press and River. He has also edited 2 books for CRC Press and River. He has participated in organizing committees, technical program committees, and as guest speaker in more than 10 conference and webinars. He has also reviewed and evaluated more than 50 papers from conferences and journals, as well as book chapters and science articles in AI, Data Science, IoT, blockchain, and Cloud Computing for CRC Press, Springer, Elsevier, Emerald, IGI-Global, and InderScience Publishers. He is an Associate Editor for the JOE, IET, Wiley, and other journals. He has taken part in workshops, webinars, and FDPs as resource person. He has more than 30 published papers, copyrights, and patents in his name. He has been awarded by international and national science societies for his eminent contributions in the R&D field. He has received travel grants and is also a member of organizations including, but not limited to IET, IEEE, ISTE, and ACM.

Seema Sharma is Assistant Professor in the Faculty of Computer Applications, Manav Rachna International Institute of Research and Studies, Faridabad, India. She is engaged in teaching MCA and BCA Classes and has teaching experience of more than 14 years. She has published more than 16 research papers in national and international journals. She has guided various MCA students in their projects. Her main research interest areas include Software Testing, Genetic Algorithm, Java, and Mathematics.

Contributors

Khalid Albulayhi
University of Idaho
Moscow, ID, USA

Qasem Abu Al-Haija
Princess Sumaya University for Technology (PSUT)
Amman, Jordan

B. Akoramurthy
Sri Venkateswara College of Engineering & Technology
Chittoor, Andhra Pradesh, India

Abdulah S. Almahayreh
Hail University
Hail, Saudi Arabia

Manivannan Babu
Bharathidasan University
Tamil Nadu, India

S. Balasubramanian
Alagappa University
Karaikudi, India

Dhruva Sreenivasa Chakravarthi
Prashanth Hospital
Vijayawada, India
and
Koneru Lakshmaiah Education Foundation Deemed to be University
Andhra Pradesh, India

Sweta Chander
Lovely Professional University
Phagwara, Punjab, India

Ruhul Amin Choudhury
Lovely Professional University
Phagwara, Punjab, India

Sarishma Dangi
Graphic Era Deemed to be University
Dehradun, India

Pushan Kumar Datta
Amity University
Kolkata, India

K. Dhivya
Pondicherry University
Kalapet, Puducherry, India.

Mohamed A. Elashiri
Beni-Suef University
Beni-Suef, Egypt

Ramya Govindaraj
Vellore Institute of Technology
Vellore, India

Shashi Kant Gupta
Integral University
Lucknow, UP, India

S. Shivam Gupta
Graphic Era Deemed to be University
Dehradun, Uttarakhand, India

Abdelwahab Said Hassan
Suez Canal University
Ismailia, Egypt

S. Hasan Hussain
Sri Venkateswara College of Engineering and Technology (Autonomous)
Chittoor, Andra Pradesh, India

K. Mohamed Jasim
Vellore Institute of Technology
Vellore, India

Chinnadurai Kathiravan
Vellore Institute of Technology
Vellore, India

Sarfraz Fayaz Khan
Algonquin College
Ottawa, Canada

Rohit Khokher
Vidya Prakashan Mandir Private
Limited
Meerut, India

Akhilesh Kumar
Gaya College
Gaya, Bihar, India

Rajendra Kumar
Sharda University
Greater Noida, Uttar Pradesh, India

Sumit Kumar
Indian Institute of Management
Kozhikode, India
and
CERES Group Inc. (Home – The
Ceres Group)
Boston, MA, USA

Nihar Ranjan Nayak
Presidency University
Bengaluru, India

Shivlal Mewada
Govt. College, Makdone (Vikram
University)
Ujjain, India

Susanta Mitra
University of the People
Pasadena, CA, USA

Sagar Dhanraj Pande
Lovely Professional University
Phagwara, Punjab, India

Ranu Pareek
Manipal University Jaipur
Rajasthan, India

M. Prasad
VIT Chennai Campus
Chennai, India

A. Shenbaga Bharatha Priya
Anna University
Chennai, India

Samrat Ray
Peter the Great Saint Petersburg
Polytechnic University
Saint Petersburg, Russia
and
Sunstone CIEM Kolkata Campus
Kolkata, India

Bhaskar Roy
Asansol Engineering College
Asansol, West Bengal, India

Sanjaya Kumar Sarangi
Utkal University
Bhubaneswar, Odisha, India

G. Vennira Selvi
Sri Venkateswara College of
Engineering and Technology
Chittoor, Andhra Pradesh, India

Sachin Sharma
Graphic Era Deemed to be
University
Dehradun, India

Mandeep Singh
Lovely Professional University
Phagwara, Punjab, India

Ram Chandra Singh
Sharda University
New Delhi, India

List of Contributors

T. B. Sivakumar
Sri Venkateswara College of
Engineering and Technology
(Autonomous)
Chittoor, Andra Pradesh, India

Parin Somani
Independent Academic Scholar
New Delhi, India

Mukesh Soni
Jagran Lakecity University
Bhopal, India

S. J. Sultanuddin
MEASI Institute of Information
Technology
Tamil Nadu, India

R. K. Tailor
Manipal University Jaipur
Jaipur, Rajasthan, India.

Ashish Kumar Tamrakar
Bhilai Institute of Technology (BIT)
Raipur Chhattisgarh, India

Sunil Kumar Vohra
Amity University Noida
Uttar Pradesh, India

1 Comprehensive Analysis of Fundamentals, Innovation, and Key Challenges of Blockchain

Mukesh Soni, Mohamed A. Elashiri, Abdulah S. Almahayreh, and Abdelwahab Said Hassan

CONTENTS

1.1	Introduction	2
1.2	Research Background	3
	1.2.1 Background of Blockchain Research	3
	1.2.2 Background of Security Vulnerability Research	3
	1.2.3 System Vulnerabilities	4
1.3	Blockchain Concept	4
1.4	Analysis of Critical Technologies of Blockchain	6
	1.4.1 Hash Algorithm	8
	1.4.2 Merkle Tree	9
1.5	Research Gap	10
	1.5.1 Data Storage and Interaction	10
	1.5.1.1 Multi-Form Data Storage	10
	1.5.2 Privacy Protection	11
	1.5.2.1 User Privacy Protection	11
	1.5.2.2 Enterprise Privacy Protection	11
	1.5.3 Resource Allocation	11
	1.5.4 Vulnerability Attack	12
	1.5.4.1 Fork Attack	12
	1.5.4.2 Cryptography-based Attacks	13
	1.5.4.3 Other Attacks	14
1.6	Conclusion	15
References		16

DOI: 10.1201/9781003269281-1

1.1 INTRODUCTION

In today's era of information explosion, it isn't easy to safely and reliably transmit massive data at any time. Therefore, data security has seriously affected people's lives. How to effectively ensure system data security has become one of the urgent issues (Frauenthaler et al., 2020).

Everyone is a producer and user of data, but companies often lead to poor data sharing and even data silos to protect privacy. Although data sharing technology can solve the problems of closeness and singleness of information and effectively realize the value-added of data, it also has the defects mentioned above of data security and privacy leakage.

In 2018, Facebook was hacked, and the information of more than 80 million users was leaked; in 2019, a large amount of user data was stolen from many companies, and hackers privately traded 870 million pieces of personal data; in 2020, the information of 5.2 million guests of Marriott International Group was leaked, etc. Criminals often use these revealed identities for illegal activities. Therefore, it is of great practical significance to study a new generation of technology to prevent hacker attacks and ensure data security.

Blockchain is a new technology developed based on cryptography, statistics, economics, and computer science. It is widely used for its advantages of decentralized storage, high anonymity, and data consistency, and to establish distributed storage and effective utilization of data in the fields of information security, finance, securities, digital right confirmation, and traceability.

The combination of blockchain and privacy protection can reduce the risk of lax third-party supervision, ensure data security and effectiveness to a certain extent, and have broader application value.

The security of the blockchain system is of great significance to the system itself. To avoid attacks by malicious nodes and make transactions proceed orderly, blockchain workers improve the consensus mechanism, optimize intelligent contracts, and strengthen network supervision.

The blockchain can apply to various fields. However, since blockchain technology is still in the development stage, it is not perfect in the core technology field, resulting in defects in the blockchain system itself and even new technical loopholes in the process of improvement. This chapter therefore mainly analyzes and summarizes some existing problems of the blockchain, outlining several key technologies of the current blockchain, and aiming to summarize existing issues in the blockchain system at this stage, and give security issues based on these critical technologies.

By elaborating on the current problems of the blockchain system: (1) it can be used as a reference for blockchain system developers to understand and avoid common pitfalls; (2) it can be used as a reference a guide for researchers to facilitate the development of analysis and verification techniques for blockchain technology.

To provide more robust theoretical and practical support for future development of blockchain and solve problems faster, therefore, analysis of the existing issues and security vulnerabilities of the blockchain system in this chapter is of great significance to the development of blockchain technology (Khang et al., AI & Blockchain, 2021).

1.2 RESEARCH BACKGROUND

1.2.1 Background of Blockchain Research

The concept of "blockchain" was first proposed in 2008 (Yang et al., 2019); it has been successfully applied in the data encryption system of the bitcoin system and in government enterprises and academic research, etc.

Focus and research hotspots: The British government released a report in January 2016 (Yang et al., 2019), pointing out that the British government is actively evaluating the potential of blockchain technology to handle leadership, collaboration and governance better; in November of the same year, Microsoft released the Azure blockchain.

In October 2016 (Davenport et al., 2019) indicated that India has officially entered an unprecedented era of rapid development of blockchain; in February of the same year, IBM launched the blockchain technology formally.

Chain application platform: In October 2019, General Secretary Xi Jinping pointed out that blockchain will become critical for national strategic deployment as an emerging independent innovation technology.

By the end of 2019, India had released 25 national standards related to cryptographic modules, and large Internet companies such as Alibaba, Tencent, JD.com; Baidu have successively launched their blockchain service platforms; and Facebook has established the Libra Association. Several states have introduced relevant policies for blockchain (Banno et al., 2019), and their government departments have proposed 22 blockchain-related bills (Dabboussi et al., 2021).

In 2020, India established a special blockchain committee. Domestic financial institutions such as China Merchants Bank, Agricultural Bank of China, and Industrial and Commercial Bank of China have launched blockchain economic projects, laying a solid foundation for promoting the rapid development of blockchain.

Blockchain technology has the characteristics of decentralized storage, privacy protection, tamper resistance, etc. It provides an open, decentralized, and fault-tolerant transaction mechanism, becoming the core of a new generation of anonymous online payment, remittance, and digital asset transactions, and is widely used in powerful trading platforms (Salman et al., 2019).

At the same time, it has also brought profound changes to finance, regulatory agencies, technological innovation, agriculture, and politics (Moschou et al., 2020). Aiming at a programmable society, blockchain technology will create a new era of technological innovation in the 21st century, just like Internet technology in the 20th century (Wu et al., 2021).

1.2.2 Background of Security Vulnerability Research

According to the National Information Security Vulnerability Sharing Platform (Noh et al., 2019), from 2016 to 2020 the distribution of low-risk, medium-risk, and high-risk vulnerabilities is compared, as shown in Figure 1.1. The frequency of medium-risk exposures is the highest. Although it decreased in 2018, it is generally increasing; the frequency of high-risk vulnerabilities is second only to medium-risk vulnerabilities as in Figure 1.1.

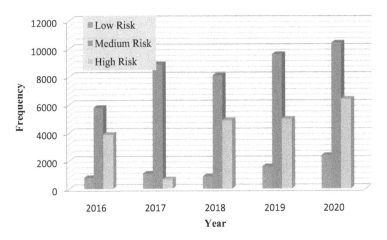

FIGURE 1.1 Vulnerability distribution.

1.2.3 System Vulnerabilities

At the same time, vulnerabilities seriously endanger the security of blockchain systems. To further analyze the hazards of blockchain vulnerabilities, the literature (Soni et al., 2021) classified them from the perspective of blockchain systems and summarized them according to the following six types of vulnerabilities.

1.3 BLOCKCHAIN CONCEPT

Since Yang et al. (2019) proposed the blockchain, scholars at home and abroad have put forward their own opinions on the concept of blockchain from different perspectives. As Sookhak mentioned, the blockchain is a decentralized, trustless, anti-interference, and distributed ledger (Kuperberg et al., 2020).

Daniel believes that blockchain is a new type of software development architecture that uses unique data structures to connect data into chains in an append-only form and uses hash functions as a tool for mapping data to ensure data security (Miyamae et al., 2021). Wang Fang defines blockchain as the use of encrypted chain block structure to verify and store data, use distributed node consensus mechanism (Consensus Mechanism) to generate and update data, and use Smart Contracts to program and store data (Khang et al., IoT & Healthcare, 2021).

Blockchain is a decentralized infrastructure and distributed computing paradigm for manipulating data (Guo et al., 2019). Yuan Yong and Zhang Ao understand the blockchain as a decentralized distributed database that grows with time series. Its essence is a distributed ledger technology based on asymmetric encryption algorithms (Li et al., 2019).

Yu describes blockchain as a peer-to-peer technical implementation of an electronic currency ledger system. Participants maintain and record each bitcoin transaction without a central server in the network system (Latifi et al., 2019).

Halamka explains that a blockchain consists of immutable distributed digital ledgers that track transactions and record them digitally.

On the block, all nodes in the blockchain architecture serve correspondingly and do not depend on the central network server, so the primary failure point of the blockchain is invulnerable (Yu et al., 2018).

Yang and Wei Xiaoxu describe blockchain as a distributed database technology that uses cryptography to connect data blocks and conduct feasible transactions. It has the functions of being tamper-proof, and enables traceability and multi-party maintenance. For information sharing and information supervision between two parties, any party must obtain the prior consent of the other parties as agreed (Zhu et al., 2019).

To sum up, the blockchain concept can be understood as based on the asymmetric encryption algorithm, with improved Merkle tree (Linoy et al., 2019) as a data structure, using consensus mechanism, peer-to-peer network, smart contract, and other technologies.

A distributed storage database technology, blockchain is divided into four categories: Public Blockchain, Consortium Blockchain, Private Blockchain, and Hybrid Blockchain (Zheng et al., 2019).

(A) The Public Chain is a blockchain system that anyone in the network can access at any time and is generally considered to be a blockchain with complete decentralization, high anonymity, and data immutability.

(B) The Consortium Chain is a blockchain jointly managed by several enterprises or institutions. Participants must register and authenticate in advance. Therefore, compared with the public chain, the consortium chain has fewer participating nodes. The data is recorded and maintained by the certified participants, and such nodes have the right to read the data.

(C) A Private Chain is a blockchain controlled by an organization or a user. The rules for controlling the number of participating nodes are strict, so the transaction speed is breakneck, the privacy level is higher, and not easily attacked. As a result, the chain system has higher security, but decentralization is significantly weakened. From the perspective of access type, blockchain is divided into permissionless and permissioned chains.

A non-licensed chain can access the blockchain system without permission. A public chain is a typical non-licensed chain, and all nodes can freely participate in transactions on the chain. Permissioned chains include alliance chains and private chains.

There are strict standards and controls for access nodes, and data access rights in the system are only authorized to authenticated nodes. Compared with the permissionless chain, the permissioned chain sacrifices a certain degree of decentralization (Saraiva et al., 2021) in exchange for higher security protection of the data on the chain. Finally, a hybrid chain is a mixture of public and private chains, combining the characteristics of public and private chains.

(D) The Hybrid Chain allows users to decide the participating blockchain members. Whether transactions can be made public, the hybrid blockchain is customizable. Hence, its hybrid architecture ensures privacy by utilizing the private blockchain's restricted access while maintaining the integrity, transparency, and security of the public blockchain.

1.4 ANALYSIS OF CRITICAL TECHNOLOGIES OF BLOCKCHAIN

Blockchain technology includes cryptography, Merkle tree, peer-to-peer (P2P), consensus mechanism, and smart contracts (Sober et al., 2021). The critical technology stack of the blockchain is shown in Figure 1.2.

The principle of cryptography in the blockchain mainly involves encryption and signature. Encryption is encrypted by the sender using the receiver's public key. After receiving the report, the receiver can decrypt the information with his private key; the sender sends a signature.

The sender signs with his private key, and the receiver uses the sender's public key to verify the identity of the sender of the information; both methods use asymmetric encryption algorithms. Merkle tree uses hash pointers to form a tree-shaped data structure and aggregates the processed data connections into a series of hash values. The algorithm consistency between nodes ensures the accuracy of hash pointers as Figure 1.2.

The transmission between control messages and data can be completed directly between nodes, which is beneficial for each node to monitor the network publishing blocks and verify information publishing transactions legality, allowing data to be stored, maintained, and distributed in a fast and secure manner (Lamken et al., 2021). The consensus mechanism ensures the joint participation of all users in the blockchain and is an essential protocol for realizing decentralized management.

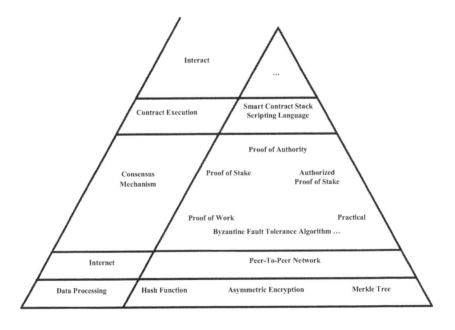

FIGURE 1.2 Key technology stack of block-chain.

Analysis of Fundamentals, Innovation, and Key Challenges

Since the public mechanism was proposed, a series of formula algorithms have been used. Smart contracts have been further improved based on blockchain, from a single-stack scripting language to a contract agreement that can automatically control the completion of transactions.

In terms of structure, a block consists of two parts: a block header (Block Header) and a block body (Block Body). The critical information in the block header includes the current version number (Version), the previous block hash value (Previews-Hash), and the timestamp (Timestamp), random number (Nonce), and Merkle tree hash value (Merkle hash) and other information (Kumar et al., 2021).

The previous block stores the hash value of the next block and connects them in the order of the time they are generated. Physically, it is the connection between blocks, and logically, it is the association of information on the chain. It is the ledger form of data association, as shown in Figure 1.3.

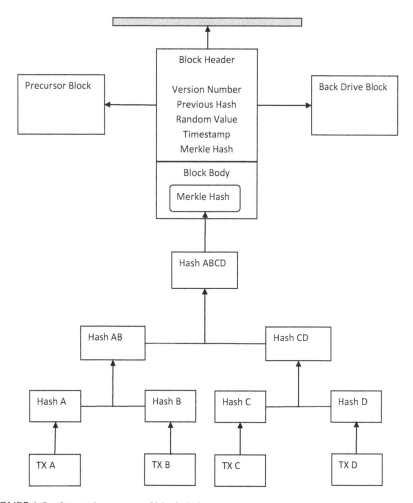

FIGURE 1.3 Internal structure of blockchain.

1.4.1 Hash Algorithm

A hash function is a function that maps variable-length data to a fixed-length digest, and any changes to the input data because of unpredictable changes in the hash column.

The SHA-256 (secure hash algorithm 256) hash function is used in the blockchain to perform a hash operation on transaction data of any length; that is, the source data is processed to obtain a string of 256-bit characters, which is convenient for unified management and storage of data, and then unified format.

The characters are packaged and stored in the block, ensuring data security to the greatest extent while reducing storage space. Hash functions have three classic properties: one-way (Hiding), collision resistance (Collision resistance), and unpredictable results (Puzzle friendly).

One-way: Knowing the input can get the output, but the known work cannot invert the information.

Collision resistance: Since the hash function has 2256 input spaces, the amount of calculation tends to be infinite, and it is almost impossible to construct an input whose result is the current value.

Unpredictable results: For published transactions or random values, it is impossible to speculate on the output of a particular feature. To ensure the security of data in the transaction process of the blockchain, as shown in the transaction signature process in Figure 1.4:

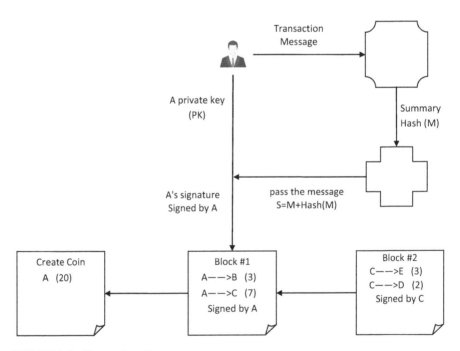

FIGURE 1.4 Transaction signature.

- first, the hash value Hash (M) is taken for all generated transaction data;
- secondly, the obtained hash value is spliced with the transaction plaintext to obtain;
- next send the content S; then digitally sign S with the sender's private key;
- and finally, write the digital signature into the block.

The digital signature uses the computational security of asymmetric algorithms to ensure the integrity and nonrepudiation of information. For example, when A initiates a transaction to B, A claims to be A, but B cannot determine the correctness of the message and A's identity because B's public key is public. Anyone can use B's public key to send messages to it.

1.4.2 Merkle Tree

Blockchain is a distributed storage technology for data based on cryptography in a broad sense. First, preprocess the data: the blockchain encrypts, hashes and digests the data to obtain a data format that meets the requirements of the block, which reduces the storage space and ensures the security of the data to the greatest extent.

The different data hash values obtained are packaged and put into the block. Merkle tree (Son et al., 2021) is a tree data structure in the computer field, as shown in Figure 1.3. The league is divided into block header and block body. The two parts are stored and connected using the data structure of the Merkle tree. The hash value is obtained by the transaction (transaction, TX) of the leaf node.

The leaf nodes of the previous layer, in this way, go up layer by layer and package the root hash value obtained by all transactions in the block body iteratively and store it in the block header. Add critical values such as transaction timestamps and random numbers when packaging transactions. Merkle trees can realize fast storage and reading of extensive data, conducive to efficient data verification.

As a critical structure for blockchain data storage, Merkle tree is mainly used to check complete transactions (Xu et al., 2019). Merkle tree divides network nodes into full nodes and light nodes. The full node stores all the information in the transaction on the chain. In contrast, the light node is responsible for verifying the transaction's legitimacy, such as a bitcoin wallet. The main steps are:

(A) The light node calculates the Hash B of TX B and sends a request to all nodes in the chain to request a Merkle proof to prove that the blue transaction TX B is included in the Merkle tree.
(B) The full node receives the request and sends the green Hash A of the previous layer to the light node for calculation and verification.
(C) The light node obtains Hash A, performs a hash operation with its Hash B to get the Hash AB value, and the entire node transmits the green Hash CD on another branch to the light node.
(D) The light node performs a hash operation on its Hash AB and the obtained Hash CD finally accepts Hash ABCD.
(E) Finally, compare the result of (4) with the root hash value stored locally to verify the existence of the transaction.

1.5 RESEARCH GAP

Blockchain is developing according to the trend, but many problems restrict and hinder its advancement. To better analyze the issues in the development of blockchain technology, this chapter will summarize the four aspects of blockchain data storage and interaction, privacy protection, resource allocation, and vulnerability attacks.

1.5.1 DATA STORAGE AND INTERACTION

Blockchain has certain advantages in data storage and reading as a distributed database. Still, with the increase of data, most nodes cannot store data efficiently, and problems such as multiform data storage, data update, and cross-chain delay occur.

1.5.1.1 Multi-Form Data Storage

One of the most significant advantages of blockchain is the one-time storage of non-relational data and multiple reads and random deployment of nodes in a dense peer-to-peer network (P2P) system for data hash storage. For legitimate users, they can inquire about their own or others' transaction amount, internal system information or block-list information, but all data is presented in the form of a string. Suppose the transaction data is video, audio or picture, with the increase of network demand. The network delay causes the data transmission to lose frames, and the distributed ledger cannot guarantee the distributed consistency.

Data update: Due to the serialization feature of the blockchain, each piece of data on the chain needs to be sorted and connected, so block transactions are extremely slow. The average bitcoin transaction time is: 10 minutes per block, about 3 to 4 transactions per second, and the Ethereum block time is about 15 seconds, about 20 transactions per second. In comparison, Visa's payment platform can process 65,000 transactions per second, and Alipay processes 89,000 transactions per second, nearly a thousand times faster than bitcoin and Ethereum.

The high transaction rate is highly unfavorable to fast access and interaction data. If the block time is shortened, the inconsistency of the data after the data update caused by the fork can be minimized.

Cross-chain delay: The blockchain combines the characteristics of decentralization and anonymity, and the encrypted digital currency based on this can be circulated the world.

As a point-to-point payment system, the operation of cross-border transactions is more convenient and faster than traditional checks and remittances. It eliminates the time-consuming manual travel, reduces the error rate caused by manual processes, and improves exchange rates between different countries. But with that comes an increase in fees.

Due to the geographical span of various countries, the data transmission delay is significant, and it is more vulnerable to malicious attacks during the transmission process. The transaction confirmation wait time is extended. Therefore, the distributed storage of the blockchain transaction ledger also needs to consider the security of data transmission: flexibility and scalability.

1.5.2 PRIVACY PROTECTION

Privacy protection refers to the process of maintaining and protecting the private information of individuals or enterprises through technical means. The following mainly divides privacy protection into two categories: user and enterprise privacy protection.

1.5.2.1 User Privacy Protection

For users, privacy protection protects the data information that users do not want to be disclosed. Anonymity isolates the blockchain from the real world. When using blockchain, it is difficult for users to keep their identities, IP addresses, contact information, public and private keys, and names unrelated. Attackers conduct transactions with a large number of users.

The address uses traffic analysis and other methods to conduct network attacks. If the account address used does not have a high level of security protection, once the private key is leaked, the blockchain will not have any change mechanism, and the transaction account can only be abandoned.

1.5.2.2 Enterprise Privacy Protection

Every process of transactions on the blockchain is open and transparent, but for enterprises, complete transparency of data does not mean all advantages.

For example, every transaction requires real-time data operations in a stock exchange. All users can see the transaction data of other nodes, which is easy to cause criminals to misappropriate user privacy and cause property damage, which makes the enterprise lose the trust of most users.

1.5.3 RESOURCE ALLOCATION

Blockchain has decentralization and tampers resistance characteristics, so the data on the ledger is prone to problems such as transaction order dependence, high data redundancy, unsustainable development, and waste of computing power for the system. Transaction ordering dependency: In Ethereum, the more ether you pay, the higher the transaction ordering priority and faster.

When two nodes submit transactions simultaneously, the transaction of the node that spends more ether will be processed first so that the node that consumes less has to wait in order. The node that pays more subsequently can join the queue to execute the transaction first without waiting, causing Blockage delays in progress, resulting in highly uneven distribution of resources.

High data redundancy: Blockchain relies on distributed ledgers. One advantage is distributed storage of data. However, with more and more data recorded, the information shows explosive growth. To ensure that the ledger is highly consistent, real-time replication, synchronization, and updating of each node result in highly redundant data, and the storage capacity of each node is limited. The future faces the dilemma of what to do with this data (Khang et al., AI-Centric Smart City Ecosystem, 2022).

Unsustainable: Paper money circulates because it is reusable, exchangeable, and market regulated. However, for virtual currency, the number of coins is limited and cannot be regulated with changes in the market economy. Due to the fixed nature of cash, it cannot be circulated as currency (Nyaletey et al., 2019).

Secondly, a transaction fee will be charged during each transaction. As the number of remaining blocks decreases, the corresponding block reward will also continue to decrease. In the future, transaction fees will become the driving force for block generation, whether suitable for economic development or technological progress.

Waste of computing power: Proof of Work is a consensus mechanism based on computer computing power to compete for the right to bookkeeping. It is expressed as a "mining" method in bitcoin, which is a natural "bounty loophole" – the people involved the more democratic, the more, the more secure the data, the higher the public participation, the problem is that while the cost is reduced, the possibility of being attacked also increases.

This powerfully exposes the drawbacks of the token mechanism; that is, the right of bookkeeping is entirely determined by the computing power, and the specialization of mining equipment drives people to spend a lot of money to buy property.

Higher ASIC chips can compete for computing power, resulting in a higher probability of obtaining accounting rights with the higher computing power of the chip. The final competition for accounting rights is proportional to the increased computing power of the chip.

The use of ASIC chips for many hash calculations results in massive energy consumption. According to statistics, during the rapid development of blockchain, the electricity used for mining every year is equivalent to the total output of several nuclear power plants.

1.5.4 VULNERABILITY ATTACK

Hackers use system vulnerabilities to monitor, intercept, replay, or even destroy them; these are called vulnerability attacks. Therefore, due to the exposure of the blockchain system itself, hacker attacks are brutal to prevent. This section divides the types of vulnerability attacks into fork attacks. Cryptography-based attacks, consensus-based attacks, clever contract-based attacks, and other attacks.

1.5.4.1 Fork Attack

Fork attacks are more common in blockchain systems and are incredibly harmful. Fork attacks are mainly divided into two categories: the system generates forks, which leads to being attacked, and the attackers actively create divisions to attack (Narayanam et al., 2021).

The separation caused by the split of the system itself: First of all, under the premise of weak consensus, the blockchain will generate multiple blocks (state fork) at the same time due to the characteristics of the system chronologically developing partnerships, so it is easy to become an attacker. Second, due to the update of the internal protocol of the blockchain system, such as software upgrade, weak consensus cannot require all nodes in the entire system to be updated at the same time, which may cause hard fork or soft fork (IEEE et al., 2021).

The old node does not recognize the new node in the hard fork, and as long as the old node does not update the system fork, it will not disappear. A soft fork is the opposite; the fork threat is eliminated as long as blocks are added along with the new node (Mitani & Otsuka, 2019).

Attackers actively create forks to attack: a Sybil attack, a type of attack based on a P2P network. It will be caused by the attack server generating more than 51% of the system's puppet accounts to participate in voting. The redundancy mechanism of distributed storage systems can be defeated, and it also poses a threat to humanity in sensor networks (IEEE et al., 2021).

The second is the double-spending attack (Aoki et al., 2021), which exploits the transmission properties of the bitcoin digital cryptocurrency to make money "spend twice" unless the recipient verifies each is ghostwriting immediately when the transaction is initiated. Otherwise, it will cause huge losses. In bitcoin, transaction signatures are used to prevent double-spending attacks, and many types of blockchain hardware will also use tamper-proof modes (Malik et al., 2019).

Third, to prevent double-spending attacks, Ethereum uses multiple data sets to generate hash values for blocks, but due to network delays, replay attacks will occur (Zonda et al., 2020); that is, the same request is reused in the Ethereum system. As a result, the money is "charged twice," which is the opposite of the "double-spend attack."

The fourth is the private mining attack (IEEE et al., 2021), which mainly refers to the malicious mining pool digging out blocks and not publishing them, but continuing to mine on the hidden chain. When the legitimate chain maintained is longer, the negative mining pool publishes the branched-chain.

Finally, the fork attack is successfully implemented since honest miners choose the longest chain as the legitimate chain.

1.5.4.2 Cryptography-based Attacks

The security protection of data in the blockchain is based on cryptography's purely mathematical calculation method. The cryptographic algorithms used include MAC function, HASH function, RSA public-key encryption system and ECC elliptic curve encryption system (Putz et al., 2020). These algorithms are safe at present, and the generation of the key requires a particular random element.

The security factor of a key pair generated by a random excellent component is higher than poor random stuff. It can directly avoid generating duplicate key pairs with other accounts.

For the cracking of critical teams, the most common password blasting method is dictionary attack: the steps of matching account passwords are automatically executed by constructing scripts of common password combination patterns, but the password combinations built by such hands are relatively single, and for a complex account, the password is impossible to start.

Passive attack (Chicaiza et al., 2021) is based on the traffic analysis of the intercepted data PDUs by the attacker and the cumulative response of the data packets; the attacker can obtain the length, frequency, and characteristics of the data and even decipher the information content.

Due to many nodes in the blockchain system, the authenticated participants can't be entirely sure whether they are honest or not.

Similarly, side-channel attacks (Robinson et al., 2021) are common in any blockchain system – the server hardware stores critical information. If the private key is stolen, the account corresponding to the private key can be locked (Cheng & Shaoqin, 2020).

The private key in the system cannot be retrieved if it is stolen, and it is impossible to regenerate the same private key in the calculation. Since blockchain data cannot be changed, once a criminal steals a private key and publicly transfers the associated funds to another account, the transaction is usually irreversible.

The input space of the current mainstream SHA-256 algorithm is $2256 \approx 1077$. Although the amount of computation required by an ordinary computer around the clock for two years, the calculation rate will increase exponentially if a quantum computer is used. Therefore, with the quantum computer (Agyekum et al., 2021), the existing cryptosystem will directly face the threat of being breached.

1.5.4.3 Other Attacks

In addition to the above four types of vulnerability attacks based on blockchain systems, there are also some vulnerability attacks based on network, social engineering, or physical devices, which attackers can use to steal or destroy data from outside the system.

For example, distributed systems allow physically separated users to interact online, and eclipse attacks can prevent targeted users from communicating with the outside world.

Virus attack: The blockchain platform is widely used at this stage, and there is no unified standard for normalization. It is easy for attackers to implant viruses or Trojans, which will seriously threaten the financial status of users (Pal et al., 2019).

Supply chain attack: Most companies will outsource their large-scale business to other technology companies or joint firms of multiple companies, but in many cases, this potential danger is the partner, who may be a rogue company or a rogue employee, there are loopholes in the system's creation.

Man-in-the-middle attack: The attacker intercepts, eavesdrops or even tampers the information transmitted in the network but does not alert both parties to the attack.

Replay attack: Combining the last transmitted information with the current transmitted details and then sending it to deceive the system and hinder authentication.

Social engineering analysis (Halevi et al., 2019): Perform traffic analysis on the transaction data of specific accounts and obtain the user's real identity by linking his life trajectory in reality.

Malware attack: The attacker defrauds the user's login information by sending malware to steal account information and node private keys.

Side-channel attack: The attacker attacks or even destroys the physical devices that support the blockchain operation under the chain, resulting in damage to the server hard disk and data loss. While the attacker achieves the integrity of the blockchain's distributed ledger, there is little benefit to the attacker itself. In addition to the types of attacks listed above, there are various attacks (Cheng et al., 2020).

1.6 CONCLUSION

With the rapid development of blockchain technology, blockchain has the advantages of decentralized data management, non-tampering and high security, so blockchain has attracted widespread attention from the government, enterprises and scholars. It has been successfully applied in some fields (Gul et al., 2020).

However, while the blockchain provides technology for decentralized platforms, the security problems caused by its system vulnerabilities are also becoming more and more serious. Based on the above issues, this chapter starts with the critical technology and vulnerability classification analysis of the blockchain and draws the following conclusions through literature analysis and comprehensive research (Jiang et al., 2019).

This chapter summarizes the research status at home and abroad and the evolution trend of vulnerabilities since blockchain development. It is found that the security problems of application programs, operating systems, and databases are still severe.

Through the analysis of research literature, the classification of current vulnerabilities in blockchain systems is obtained. Then it summarizes the concept of blockchain and conducts an in-depth analysis of crucial blockchain technologies such as cryptography, Merkle tree, peer-to-peer (P2P), consensus mechanism, and smart contracts, and summarizes their working principles and their advantages (Kuzlu et al., 2019).

In particular, it compares and analyzes the application effects of mainstream blockchain consensus algorithms on many platforms and summarizes the appropriate application scenarios of various key technologies to facilitate users' reference.

This chapter analyzes and summarizes the four aspects of blockchain data storage and interaction, privacy protection, resource allocation, and vulnerability attacks, and summarizes and points out the problems that need to be solved urgently in the security of blockchain systems (Fedorov et al., 2021).

The vulnerability attacks are outlined, and five types of exploitable vulnerability attacks are pointed out, including fork attacks, cryptography-based attacks, consensus algorithm-based attacks, clever contract-based attacks, and attacks launched from outside the system (Guo et al., 2020).

According to the classification of security vulnerability attacks, the definition of each attack category and possible effective solutions are obtained from many kinds of literature, which point out the direction for users to improve the security of the blockchain. The above four problems pose a significant security threat to the blockchain system and must be paid attention to and guarded against (Harris & Waggoner, 2019).

Combining the above problems, this chapter suggests that the security system needs to be more perfected in future system design. The security of the blockchain system should be improved from the algorithm to the system as a whole. Among them, the storage efficiency of data needs to be further enhanced, and the transaction volume in milliseconds (Faria & Correia, 2019).

The number directly affects the performance of the blockchain system, so it is necessary to provide a more efficient consensus mechanism, the protection of private data needs to increase the function of access control to prevent malicious intrusion by third parties, and the allocation of resources by the system can use cloud servers or IPFS (Bhambri et al., Cloud & IoT, 2022).

Therefore, in the future, it is necessary to improve the internal design of the blockchain further to provide the Chain technology development and platform development provide more accurate (Khang et al., Data-Driven Blockchain Ecosystem, 2022), secure and standard technologies to prevent various vulnerabilities from attacking the blockchain system, thereby improving the work efficiency of the blockchain system (Gomathi et al., 2021).

REFERENCES

Agyekum, K. O. -B. O., Q. Xia, E. B. Sifah, C. N. A. Cobblah, H. Xia, & J. Gao, A proxy re-encryption approach to secure data sharing in the Internet of Things based on blockchain, *IEEE Systems Journal*, 2021, doi: 10.1109/JSYST.2021.3076759

Aoki, Y. & K. Shudo, Proximity neighbor selection in blockchain networks, *2019 IEEE International Conference on Blockchain (Blockchain)*, 2019, pp. 52–8, doi: 10.1109/Blockchain.2019.00016

Banno, R., & K. Shudo, Simulating a blockchain network with SimBlock, *2019 IEEE International Conference on Blockchain and Cryptocurrency (ICBC)*, 2019, pp. 3–4, doi: 10.1109/BLOC.2019.8751431

Baranwal Somy, N., et al., Ownership preserving AI market places using blockchain, *2019 IEEE International Conference on Blockchain (Blockchain)*, 2019, pp. 156–65, doi: 10.1109/Blockchain.2019.00029

Baza, M., M. Nabil, M. Ismail, M. Mahmoud, E. Serpedin, & M. Ashiqur Rahman, Blockchain-based charging coordination mechanism for smart grid energy storage units, *2019 IEEE International Conference on Blockchain (Blockchain)*, 2019, pp. 504–9, doi: 10.1109/Blockchain.2019.00076

Bhambri, P., S. Rani, G. Gupta, & Khang A., *Cloud and Fog Computing Platforms for Internet of Things*. Chapman & Hall. ISBN: 9781032101507, 2022, doi: 10.1201/9781032101507

Cheng, Y., Music information retrieval technology: Fusion of music, artificial intelligence and blockchain, *2020 3rd International Conference on Smart BlockChain (SmartBlock)*, 2020, pp. 143–6, doi: 10.1109/SmartBlock52591.2020.00033

Cheng, Y., & H. Shaoqin, Research on blockchain technology in cryptographic exploration, *2020 International Conference on Big Data & Artificial Intelligence & Software Engineering (ICBASE)*, 2020, pp. 120–3, doi: 10.1109/ICBASE51474.2020.00033

Chicaiza, S. A. Y., C. N. S. Chafla, L. F. E. Álvarez, P. F. I. Matute, & R. D. Rodriguez, Analysis of information security in the PoW (Proof of Work) and PoS (Proof of Stake) blockchain protocols as an alternative for handling confidential nformation in the public finance ecuadorian sector, *2021 16th Iberian Conference on Information Systems and Technologies (CISTI)*, 2021, pp. 1–5, doi: 10.23919/CISTI52073.2021.9476382

Dabboussi, S., F. Victor, & W. Prinz, BCDM – A decision and operation model for blockchains, *2021 IEEE International Conference on Blockchain and Cryptocurrency (ICBC)*, 2021, pp. 1–3, doi: 10.1109/ICBC51069.2021.9461146

Davenport, A., & S. Shetty, Air gapped wallet schemes and private key leakage in permissioned blockchain platforms, *2019 IEEE International Conference on Blockchain (Blockchain)*, 2019, pp. 541–5, doi: 10.1109/Blockchain.2019.00004

Faria, C., & M. Correia, BlockSim: Blockchain simulator, *2019 IEEE International Conference on Blockchain (Blockchain)*, 2019, pp. 439–46, doi: 10.1109/Blockchain.2019.00067

Fedorov, I. R, A. V. Pimenov, G. A. Panin, & S. V. Bezzateev, Blockchain in 5G networks: Perfomance evaluation of private blockchain, *2021 Wave Electronics and its Application in Information and Telecommunication Systems (WECONF)*, 2021, pp. 1–4, doi: 10.1109/WECONF51603.2021.9470519

Frauenthaler, P., M. Sigwart, C. Spanring, M. Sober, & S. Schulte, ETH relay: A costefficient relay for Ethereum-based blockchains, *2020 IEEE International Conference on Blockchain (Blockchain)*, 2020, pp. 204–13, doi: 10.1109/Blockchain50366.2020.00032

Gomathi, S., M. Soni, G. Dhiman, R. Govindaraj, & P. Kumar, A survey on applications and security issues of blockchain technology in business sectors, *Materials Today: Proceedings*, 2021, ISSN 2214–7853.

Gul, M. J. J., A. Paul, S. Rho, & M. Kim, Blockchain based healthcare system with artificial intelligence, *2020 International Conference on Computational Science and Computational Intelligence (CSCI)*, 2020, pp. 740–1, doi: 10.1109/CSCI51800.2020.00138

Guo, H., W. Li, M. Nejad, & C. -C.Shen, Access control for electronic health records with hybrid blockchain-edge architecture, *2019 IEEE International Conference on Blockchain (Blockchain)*, 2019, pp. 44–51, doi: 10.1109/Blockchain.2019.00015

Guo, X., Q. Guo, M. Liu, Y. Wang, Y. Ma, & B. Yang, A certificateless consortium blockchain for IoTs, *2020 IEEE 40th International Conference on Distributed Computing Systems (ICDCS)*, 2020, pp. 496–506, doi: 10.1109/ICDCS47774.2020.00054

Halevi, T., F. Benhamouda, A. De Caro, C. Jutla, Y. Manevich, & Q. Zhang, Initial public offering (IPO) on permissioned blockchain using secure multiparty computation, *2019 IEEE International Conference on Blockchain (Blockchain)*, 2019, pp. 91–8, doi: 10.1109/Blockchain.2019.00021

Harris, J. D., & B. Waggoner, Decentralized and collaborative AI on blockchain, *2019 IEEE International Conference on Blockchain (Blockchain)*, 2019, pp. 368–75, doi: 10.1109/Blockchain.2019.00057

Jiang, S., J. Cao, J. A. McCann, Y. Yang, Y. Liu, X. Wang, & Y. Deng, Privacy-preserving and efficient multi-keyword search over encrypted data on blockchain, *2019 IEEE International Conference on Blockchain (Blockchain)*, 2019, pp. 405–10, doi: 10.1109/Blockchain.2019.00062

Khang, A., Geeta Rana, Ravindra Sharma, Alok Kumar Goel, & Ashok Kumar Dubey, The role of artificial intelligence in blockchain applications, *Reinventing Manufacturing and Business Processes Through Artificial Intelligence*, 19–38, 2021, https://doi.org/10.1201/9781003145011

Khang, A., S. Rani, M. Chauhan, & A. Kataria, IoT equipped intelligent distributed framework for smart healthcare systems, *Networking and Internet Architecture*, 2021, https://arxiv.org/abs/2110.04997v2, doi: 10.48550/arXiv.2110.04997

Khang, A., Sita Rani, & Arun Kumar Sivaraman, *AI-Centric Smart City Ecosystem: Technologies, Design and Implementation*, HB.ISBN: 978-1-032-17079-4 ** EB.ISBN: 978-1-003-25254-2, 2022, doi: 10.1201/9781003252542

Kumar, R., B. Palanisamy, & S. Sural, BEAAS: Blockchain enabled attribute-based access control as a service, *2021 IEEE International Conference on Blockchain and Cryptocurrency (ICBC)*, 2021, pp. 1–3, doi: 10.1109/ICBC51069.2021.9461151

Kuperberg, M., Towards enabling deletion in append-only blockchains to support data growth management and GDPR compliance, *2020 IEEE International Conference on Blockchain (Blockchain)*, 2020, pp. 393–400, doi: 10.1109/Blockchain50366.2020.00057

Kuzlu, M., M. Pipattanasomporn, L. Gurses, & S. Rahman, Performance analysis of a hyperledger fabric blockchain framework: Throughput, latency and scalability, *2019 IEEE International Conference on Blockchain (Blockchain)*, 2019, pp. 536–40, doi: 10.1109/Blockchain.2019.00003

Lamken, D., T. Wagner, T. Hoiss, K. Seidenfad, A. Hermann, M. Kus, & U. Lechner, Design patterns and framework for blockchain integration in supply chains, *2021 IEEE International Conference on Blockchain and Cryptocurrency (ICBC)*, 2021, pp. 1–3, doi: 10.1109/ICBC51069.2021.9461062

Latifi, S., Y. Zhang, & L. -C. Cheng, Blockchain-based real estate market: One method for applying blockchain technology in commercial real estate market, *2019 IEEE International Conference on Blockchain (Blockchain)*, 2019, pp. 528–35, doi: 10.1109/Blockchain.2019.00002

Li, L., A. Arab, J. Liu, J. Liu, & Z. Han, Bitcoin options pricing using LSTM-based prediction model and blockchain statistics, *2019 IEEE International Conference on Blockchain (Blockchain)*, 2019, pp. 67–74, doi: 10.1109/Blockchain.2019.00018

Li, S., H. Xiao, H. Wang, T. Wang, J. Qiao, & S. Liu, Blockchain dividing based on node community clustering in intelligent manufacturing CPS, *2019 IEEE International Conference on Blockchain (Blockchain)*, 2019, pp. 124–31, doi: 10.1109/Blockchain.2019.00025

Linoy, S., H. Mahdikhani, S. Ray, R. Lu, N. Stakhanova, & A. Ghorbani, Scalable privacy-preserving query processing over Ethereum blockchain, *2019 IEEE International Conference on Blockchain (Blockchain)*, 2019, pp. 398–404, doi: 10.1109/Blockchain.2019.00061

Malik, S., V. Dedeoglu, S. S. Kanhere, & R. Jurdak, TrustChain: Trust management in blockchain and IoT supported supply chains, *2019 IEEE International Conference on Blockchain (Blockchain)*, 2019, pp. 184–93, doi: 10.1109/Blockchain.2019.00032

Mitani, T., & A. Otsuka, Traceability in permissioned blockchain, *2019 IEEE International Conference on Blockchain (Blockchain)*, 2019, pp. 286–93, doi: 10.1109/Blockchain.2019.00045

Miyamae, T., F. Kozakura, M. Nakamura, S. Zhang, S. Hua, B. Pi, & M. Morinaga, ZGridBC: Zero-knowledge proof based scalable and private blockchain platform for smart grid, *2021 IEEE International Conference on Blockchain and Cryptocurrency (ICBC)*, 2021, pp. 1–3, doi: 10.1109/ICBC51069.2021.9461122

Moschou, K. et al., Performance evaluation of different hyperledger sawtooth transaction processors for blockchain log storage with varying workloads, *2020 IEEE International Conference on Blockchain (Blockchain)*, 2020, pp. 476–81, doi: 10.1109/Blockchain50366.2020.00069

Narayanam, K., S. Goel, A. Singh, Y. Shrinivasan, & P. Selvam, Blockchain based accounts payable platform for goods trade, *2021 IEEE International Conference on Blockchain and Cryptocurrency (ICBC)*, 2021, pp. 1–5, doi: 10.1109/ICBC51069.2021.9461053

Noh, S., & K. -H. Rhee, Implicit Authentication in neural key exchange based on the randomization of the public blockchain, *2020 IEEE International Conference on Blockchain (Blockchain)*, 2020, pp. 545–9, doi: 10.1109/Blockchain50366.2020.00079

Nyaletey, E., R. M. Parizi, Q. Zhang, & K. -K. R. Choo, BlockIPFS – blockchain enabled interplanetary file system for forensic and trusted data traceability, *2019 IEEE International Conference on Blockchain (Blockchain)*, 2019, pp. 18–25, doi: 10.1109/Blockchain.2019.00012

Pal, P., & S. Ruj, BlockV: A blockchain enabled peer-peer ride sharing service, *2019 IEEE International Conference on Blockchain (Blockchain)*, 2019, pp. 463–8, doi: 10.1109/Blockchain.2019.00070

Putz, B., & G. Pernul, Detecting blockchain security threats, *2020 IEEE International Conference on Blockchain (Blockchain)*, 2020, pp. 313–20, doi: 10.1109/Blockchain 50366.2020.00046

Robinson, P., & R. Ramesh, General purpose atomic crosschain transactions, *2021 IEEE International Conference on Blockchain and Cryptocurrency (ICBC)*, 2021, pp. 1–3, doi: 10.1109/ICBC51069.2021.9461132

Salman, T., R. Jain, & L. Gupta, A reputation management framework for knowledge-based and probabilistic blockchains, *2019 IEEE International Conference on Blockchain (Blockchain)*, 2019, pp. 520–7, doi: 10.1109/Blockchain.2019.00078

Saraiva, R., A. A. Araújo, P. Soares, & J. Souza, MIRIAM: A blockchain-based web application for managing professional registrations of medical doctors in Brazil, *2021 IEEE International Conference on Blockchain and Cryptocurrency (ICBC)*, 2021, pp. 1–2, doi: 10.1109/ICBC51069.2021.9461051

Sober, M., G. Scaffino, C. Spanring, & S. Schulte, A voting-based blockchain interoperability oracle, *2021 IEEE International Conference on Blockchain (Blockchain)*, 2021, pp. 160–9, doi: 10.1109/Blockchain53845.2021.00030

Son, D., S. Al Zahr, & G. Memmi, Performance analysis of an energy trading platform using the Ethereum blockchain, *2021 IEEE International Conference on Blockchain and Cryptocurrency (ICBC)*, 2021, pp. 1–3, doi: 10.1109/ICBC51069.2021.9461115

Soni, M., & D. K. Singh, Blockchain implementation for privacy preserving and securing the healthcare data, *2021 10th IEEE International Conference on Communication Systems and Network Technologies (CSNT)*, 2021, pp. 729–34, doi: 10.1109/CSNT51715.2021.9509722

Wu, H., L. Li, H. -y. Paik, & S. S. Kanhere, MB-EHR: A multilayer blockchain-based EHR, *2021 IEEE International Conference on Blockchain and Cryptocurrency (ICBC)*, 2021, pp. 1–3, doi: 10.1109/ICBC51069.2021.9461075

Xu, R., S. Y. Nikouei, Y. Chen, E. Blasch, & A. Aved, BlendMAS: A blockchain enabled decentralized microservices architecture for smart public safety, *2019 IEEE International Conference on Blockchain (Blockchain)*, 2019, pp. 564–71, doi: 10.1109/Blockchain.2019.00082

Yang, S., Z. Chen, L. Cui, M. Xu, Z. Ming, & K. Xu, CoDAG: An efficient and compacted DAG-based blockchain protocol, *2019 IEEE International Conference on Blockchain (Blockchain)*, 2019, pp. 314–18, doi: 10.1109/Blockchain.2019.00049

Zheng, W., Z. Zheng, X. Chen, K. Dai, P. Li, & R. Chen, NutBaaS: A blockchain-as-a service platform, in *IEEE Access*, vol. 7, pp. 134422–33, 2019, doi: 10.1109/ACCESS.2019.2941905

Zhu, S., H. Hu, Y. Li, & W. Li, Hybrid blockchain design for privacy preserving crowdsourcing platform, *2019 IEEE International Conference on Blockchain (Blockchain)*, 2019, pp. 26–33, doi: 10.1109/Blockchain.2019.00013

Zonda, D., & M. Meddeb, Proxy re-encryption for privacy enhancement in blockchain: Carpooling use case, *2020 IEEE International Conference on Blockchain (Blockchain)*, 2020, pp. 482–9, doi: 10.1109/Blockchain50366.2020.00070

2 Cryptocurrency Methodologies and Techniques

S. Hasan Hussain, T. B. Sivakumar, and Alex Khang

CONTENTS

2.1 Introduction 21
2.2 Literature Survey 22
2.3 How Does the Cryptocurrency Works 23
    2.3.1 Working Method of Cryptocurrency 23
    2.3.2 Cryptocurrency Examples 24
        2.3.2.1 Bitcoin 24
        2.3.2.2 Ethereum 24
        2.3.2.3 Litecoin 24
        2.3.2.4 Ripple 24
    2.3.3 How to Buy Cryptocurrency 24
        2.3.3.1 The First Step is to Select a Platform 24
        2.3.3.2 The Seconds Step is Adding Money to Your Account 25
        2.3.3.3 The Third Step is Making a Purchase 25
    2.3.4 How to Store a Cryptocurrency 26
    2.3.5 What Can You Buy with Cryptocurrency? 26
        2.3.5.1 Websites that Deal with Technology and E-commerce 26
    2.3.6 Cryptocurrency Fraud and Cryptocurrency Scams 26
    2.3.7 Is Cryptocurrency Being Safe? 27
    2.3.8 Four Tips to Investing in Cryptocurrency Safely 28
        2.3.8.1 Research Collaborations 28
        2.3.8.2 Understand How to Keep Your Digital Currency Safe 28
        2.3.8.3 Diversify Your Portfolio 28
        2.3.8.4 Prepare for Turbulence 28
2.4 Conclusion and Future Work 28
References 29

2.1 INTRODUCTION

A cryptocurrency is a digital or virtual currency that is protected by encryption, making counterfeiting and double-spending practically impossible. Many cryptocurrencies are built on blockchain technology, which is a distributed ledger enforced by a distributed network of computers.

Cryptocurrencies are distinguished by the fact that they are not issued by any central authority, making them potentially impervious to government intervention or manipulation.

Cryptocurrencies are digital or virtual currencies that rely on cryptography technologies to function. They make it possible to make safe online payments without the involvement of third-party payment processors.

Various encryption methods and cryptographic approaches, such as elliptical curve encryption, public–private key pairs, and hashing functions, are referred to as "crypto."

2.2 LITERATURE SURVEY

The relevance of blockchain is demonstrated by the number of cryptocurrencies, which has surpassed 1900 and is still expanding. Because of the diversity of bitcoin applications, such rapid growth may soon cause interoperability issues (Tschorsch and Scheuermann, 2016).

Furthermore, as blockchain is employed in industries other than cryptocurrencies, the environment is fast changing, with Smart Contracts (SCs) playing a key role (Khang et al., AI & Blockchain, 2021). SCs, which Szabo defined in 1994 as "a computerized transaction protocol that executes the provisions of a contract" (Szabo, 1998), allow us to convert contractual clauses into embeddable code (Szabo, 2005), reducing external participation and risk.

So, a SC is an agreement between parties that, notwithstanding their lack of confidence for one another, the provisions of the agreement are automatically implemented.

As a result, blockchain technology is gaining in popularity. Almost 1000 C-suite executives (33%) said they are evaluating blockchains or have previously been actively involved with them.

Researchers and developers are already aware of the new technology's possibilities and are investigating various applications in a wide range of industries. Three generations of blockchains can be separated based on the intended audience.

While various studies of blockchain technology have been published, we believe that the state-of-the-art of blockchain-enabled applications has gotten little attention. Even in Khang et al. (AI & Blockchain, 2021), the applications and applicability of blockchains are not fully explored.

There are some reviews that focus on the specific role of blockchain (Khang et al., AI & Blockchain, 2021), such as the development of decentralized and data-intensive IoT applications, and the decentralized management of big data (Khang et al., Cloud &IoT, 2022)

Other studies look at the blockchain's security challenges, as well as its potential to promote trust and decentralization in service systems and P2P platforms. There have been investigations on some technical aspects of the blockchain design, such as its consensus protocol, the vulnerabilities of SCs, and other technical characteristics, such as its size and bandwidth, usability, data integrity, and scalability.

Cryptocurrency Methodologies and Techniques

Other studies, such as Bonneau et al., 2015, focus on the monetary component of blockchains and the security and privacy they provide.

2.3 HOW DOES THE CRYPTOCURRENCY WORKS

Cryptocurrency, often known as crypto-currency or crypto, is any type of digital or virtual currency that uses encryption to safeguard transactions. Cryptocurrencies operate without a central issuing or regulating authority, instead relying on a decentralized system to track transactions and create new units.

Cryptocurrency is a digital payment mechanism that does not rely on banks for transaction verification. It's a peer-to-peer system that allows anyone to make and receive payments from anywhere.

Cryptocurrency payments exist solely as digital entries to an online database identifying specific transactions, rather than as tangible money carried around and exchanged in the real world. The transactions that you make with cryptocurrency funds are recorded in a public ledger. Digital wallets are used to store cryptocurrency.

The moniker "cryptocurrency" comes from the fact that it uses encryption to verify transactions. This means that storing and sending cryptocurrency data between wallets and to public ledgers requires complex coding. Encryption's goal is to ensure security and safety as Figure 2.1.

2.3.1 Working Method of Cryptocurrency

Cryptocurrencies are based on the blockchain, a distributed public database that keeps track of all transactions and is updated by currency holders.

FIGURE 2.1 How does the cryptocurrency work?

Cryptocurrency units are formed through a process known as mining, which entails employing computer processing power to solve complex mathematical problems in order to earn coins. Users can also purchase the currencies from brokers, which they can then store and spend using encrypted wallets.

You don't possess anything concrete if you own cryptocurrency. What you possess is a key that enables you to transfer a record or a unit of measurement from one person to another without the involvement of a trustworthy third party.

Although bitcoin has been present since 2009, cryptocurrencies and blockchain technologies are still in their infancy in terms of financial applications, with more to come in the future. Bonds, stocks, and other financial assets might all be traded via the technology in the future.

2.3.2 Cryptocurrency Examples

Thousands of cryptocurrencies exist. Among the most well-known are:

2.3.2.1 Bitcoin

Bitcoin was the first cryptocurrency, and it is still the most widely traded, having been launched in 2009. Satoshi Nakamoto – largely assumed to be a pseudonym for an individual or group of people whose true identity is unknown – created the currency.

2.3.2.2 Ethereum

Ethereum is a blockchain platform that has its own cryptocurrency, Ether (ETH) or Ethereum. It was created in 2015. After bitcoin, it is the most widely used cryptocurrency.

2.3.2.3 Litecoin

This money is quite similar to bitcoin, but it has moved faster to build new innovations, such as speedier payments and processes that allow for more transactions.

2.3.2.4 Ripple

Ripple was founded in 2012 as a distributed ledger technology. Not only can Ripple be used to track cryptocurrency transactions, but it can also be used to track other types of transactions. Its creators have collaborated with a number of banks and financial institutions.

2.3.3 How to Buy Cryptocurrency

2.3.3.1 The First Step is to Select a Platform

The first step is to choose a platform to work with. In general, you have the option of using a regular broker or a cryptocurrency exchange:

Brokers who work in the traditional sense. These are online brokers that allow you to purchase and sell cryptocurrencies as well as other financial assets such as stocks,

Cryptocurrency Methodologies and Techniques

bonds, and exchange-traded funds (ETFs). These platforms are known for having reduced trading fees but fewer crypto features.

Exchanges for cryptocurrencies. There are a variety of cryptocurrency exchanges to choose from, each with its own set of cryptocurrencies, wallet storage options, interest-bearing account options, and other features. Asset-based fees are charged by several exchanges.

Consider which cryptocurrencies are available, the fees they charge, their security features, storage and withdrawal choices, and any educational materials when evaluating different platforms (Pankaj Bhambri et al., Cloud & IoT, 2022).

2.3.3.2 The Seconds Step is Adding Money to Your Account

After you've decided on a platform, you'll need to fund your account before you can start trading. Although this varies by platform, most crypto exchanges allow users to buy crypto with fiat (government-issued) currencies such as the US Dollar, the British Pound, or the Euro using their debit or credit cards (Alam & Zameni, 2019).

ACH and wire transfers are also accepted by some sites. The payment methods that are accepted and the time it takes to deposit or withdraw money vary each platform. Likewise, the time it takes for deposits to clear varies depending on the payment type. Fees are an essential consideration. These fees could include transaction fees for deposits and withdrawals, as well as trading fees. Fees will vary depending on the payment method and platform, so do your homework ahead of time.

Credit card purchases of cryptocurrency are deemed dangerous, and some exchanges do not allow them. Crypto transactions are also not permitted by some credit card companies. This is because cryptocurrencies are extremely volatile, and risking going into debt – or perhaps paying hefty credit card transaction fees – for particular assets is not recommended.

2.3.3.3 The Third Step is Making a Purchase

You can use the web or mobile platform of your broker or exchange to make an order. If you wish to buy cryptocurrencies, go to "buy," select the order type, enter the number of coins you want to buy, and confirm the order. Orders to "sell" follow the same procedure.

There are other ways to invest in cryptocurrency as well. PayPal, Cash App, and Venmo are examples of payment platforms that allow customers to buy, trade, or store cryptocurrencies (Afzal & Asif, 2019).

In addition, the following investment vehicles are available:

Bitcoin trusts: Bitcoin trusts can be purchased with a conventional brokerage account. Through the stock market, these vehicles provide regular investors with access to cryptocurrency.

Bitcoin mutual funds: You can select between bitcoin ETFs and bitcoin mutual funds. Blockchain stocks or ETFs: blockchain companies that specialize in the technology behind crypto and crypto transactions are another way to indirectly invest in crypto. Alternatively, you might invest in blockchain-related equities or exchange traded funds (ETFs). Your best selection will be determined by your investment objectives and risk tolerance.

2.3.4 How to Store a Cryptocurrency

Once you've purchased bitcoin, you'll need to keep it safe to avoid being hacked or stolen. Cryptocurrencies are typically stored in crypto wallets, which are physical hardware or online software that securely store the private keys to your cryptocurrencies. Some exchanges offer wallet services, allowing you to store your funds directly on the platform. However, not all exchanges or brokers will automatically give you with a wallet.

There are a variety of wallet providers from which to pick. The terms "hot wallet" and "cold wallet" are used to describe two types of wallets:

Hot wallet storage: "Hot wallets" refers to cryptocurrency storage that use internet software to safeguard your assets' private keys.
Cold wallet storage: Unlike hot wallets, cold wallets (also known as hardware wallets) save your private keys on offline electronic devices.

2.3.5 What Can You Buy with Cryptocurrency?

Bitcoin was designed from the start to be a daily transactional currency, allowing users to buy everything from a cup of coffee to a computer, as well as big-ticket things like real estate.

That hasn't happened yet, and while the number of institutions adopting cryptocurrencies is increasing, major transactions involving cryptocurrencies are still uncommon. Despite this, crypto can be used to purchase a wide range of things from e-commerce platforms. Some instances are as follows:

2.3.5.1 Websites that Deal with Technology and E-commerce

Several tech companies, like newegg.com, AT&T, and Microsoft, accept cryptocurrency on their websites. Overstock, an online retailer, was one of the first to take bitcoin. It's also accepted by Shopify, Rakuten, and Home Depot.

2.3.5.1.1 Cars
Some vehicle dealerships now accept cryptocurrencies as payment, ranging from mass-market brands to high-end luxury brands.

2.3.5.1.2 Insurance
AXA, a Swiss insurer, stated in April 2021 that it has begun taking bitcoin as a form of payment for all of its insurance lines excluding life insurance (due to regulatory issues).

In addition to accepting bitcoin for premium payments, Premier Shield Insurance, which sells house and vehicle insurance plans in the United States, also takes bitcoin for premium payments.

2.3.6 Cryptocurrency Fraud and Cryptocurrency Scams

Cryptocurrency criminality is, unfortunately, on the rise. Scams involving cryptocurrency include:

Phony websites: Scam sites with fake testimonials and crypto jargon that promise huge, guaranteed profits if you keep investing.

Virtual Ponzi schemes: Cryptocurrency thieves offer fictitious opportunities to invest in digital currencies and create the illusion of high returns by repaying old investors with money from new investors. Before its offenders were charged in December 2019, one scam organization, BitClub Network, had raised more than $700 million.

"Celebrity" endorsements: Scammers act as millionaires or well-known figures on the internet, promising to multiply your virtual currency investment but instead stealing what you contribute. They could even use messaging applications or chat forums to spread rumors that a well-known businessperson is supporting a particular cryptocurrency. The crooks sell their ownership after encouraging investors to buy and driving up the price, and the currency loses value (Duque, 2020).

Scams involving virtual currencies: The FBI has issued a warning about a new trend in online dating scams, in which con artists persuade people they meet on dating apps or social media to invest or trade in virtual currencies. In the first seven months of 2021, the FBI's Internet Crime Complaint Centre received over 1,800 reports of crypto-focused romance scams, with losses totaling $133 million.

Otherwise, fraudsters may impersonate legal virtual currency dealers or set up phony exchanges to defraud customers. Another type of crypto scam involves deceptive sales pitches for cryptocurrency-based individual retirement plans. Then there's plain cryptocurrency hacking, in which hackers gain access to people's digital wallets and steal their virtual currency.

2.3.7 Is Cryptocurrency Being Safe?

Blockchain technology is commonly used to create cryptocurrencies. The method transactions are recorded in "blocks" and time stamped is described by blockchain. It's a lengthy, complicated procedure, but the end result is a secure digital ledger of cryptocurrency transactions that hackers can't alter.

Transactions also necessitate a two-factor authentication process. To begin a transaction, you might be requested to enter a login and password. Then you may be required to input an authentication code sent to your personal cell phone through text message.

While security measures are in place, this does not mean that cryptocurrencies are impenetrable to hackers. Several high-profile thefts have wreaked havoc on bitcoin startups. Coincheck was hacked for $534 million, and BitGrail was hacked for $195 million, making them two of the most expensive cryptocurrency attacks of 2018.

The value of virtual currencies, unlike government-backed money, is solely determined by supply and demand. This can lead to dramatic swings in the market, resulting in significant gains or losses for investors.

Furthermore, compared to traditional financial instruments such as equities, bonds, and mutual funds, cryptocurrency investments are subject to significantly less governmental oversight.

2.3.8 Four Tips to Investing in Cryptocurrency Safely

All investments, according to Consumer Reports, include risk, but some experts believe bitcoin is one of the riskier investing options available. If you're thinking about investing in cryptocurrencies, these pointers can assist you in making informed decisions (Adhami et al., 2018).

2.3.8.1 Research Collaborations

Learn about bitcoin exchanges before you invest. There are around 500 exchanges to choose from, according to estimates. Before making a decision, do your homework, study reviews, and speak with more experienced investors.

2.3.8.2 Understand How to Keep Your Digital Currency Safe

You must store cryptocurrency if you purchase it. You can save it in a digital wallet or on an exchange. While there are various types of wallets, each has its own set of advantages, technological needs, and security features. You should research your storage options before investing, just as you would with exchanges (Alonso & Luis, 2019).

2.3.8.3 Diversify Your Portfolio

Diversification is essential to any successful investment strategy, and this is especially true when it comes to cryptocurrency. Don't put all of your money in bitcoin just because it's a well-known name. There are thousands of possibilities, and it's best to diversify your portfolio by investing in other currencies (Alonso-Monsalve et al., 2020).

2.3.8.4 Prepare for Turbulence

Be aware that the cryptocurrency market is quite volatile, so expect ups and downs. Prices will fluctuate dramatically. Cryptocurrency may not be a good fit for you if your investment portfolio or mental health can't manage it (Khang et al., IoT & Healthcare, 2021).

Cryptocurrency is currently all the rage, but keep in mind that it is still in its infancy and is regarded highly speculative. Be prepared for the hardships that come with investing in anything new. If you decide to participate, do your homework first and start with a little investment.

Using a comprehensive antivirus is one of the greatest methods to keep secure online. Kaspersky Internet Security protects you from malware infections, spyware, and data theft, as well as bank-grade encryption for online payments (Khang et al., AI-Centric Smart City Ecosystem, 2022).

2.4 CONCLUSION AND FUTURE WORK

In this chapter I have elaborated working method of cryptocurrency, some of the examples of cryptocurrency, how and where to buy a cryptocurrency, what are the methods are used to store the cryptocurrency and four important tips to invest the cryptocurrency.

In future, I am going to implement a novel method to provide security for cryptography.

REFERENCES

Adhami, S., Giudici, G., & Martinazzi, S. (2018). Why do businesses go crypto? An empirical analysis of initial coin offerings. *Journal of Economics and Business*, 100, 64–75. https://doi.org/10.1016/j.jeconbus.2018.04.001

Afzal, A., & Asif, A. (2019). Cryptocurrencies, blockchain and regulation: A review. *The Lahore Journal of Economics*, 24(1), 103–30.

Alam, N., & Zameni, A. P. (2019). Existing regulatory frameworks of cryptocurrency and the Shari'ah alternative. In Billah, M (eds). *Halal Cryptocurrency Management*, 179–94. London: Palgrave Macmillan.

Alonso, N., & Luis, S. (2019). Activities and operations with cryptocurrencies and their taxation implications: The Spanish case. *Laws*, 8(3), 1–13.

Alonso-Monsalve, S., Suárez-Cetrulo, A. L., Cervantes, A., & Quintana, D. (2020). Convolution on neural networks for high-frequency trend prediction of cryptocurrency exchange rates using technical indicators. *Expert Systems with Applications*, 149, 113250. https://doi.org/10.1016/j.eswa.2020.113250.

Bhambri, P., Rani, S., Gupta, G., & Khang, A. (2022). *Cloud and Fog Computing Platforms for Internet of Things*. Chapman & Hall. ISBN: 9781032101507, doi: 10.1201

Bonneau, J., Miller, A., Clark, J., Narayanan, A., Kroll, J. A., & Felten, E. W. (2015). SoK: Research perspectives and challenges for bitcoin and cryptocurrencies. *2015 IEEE Symposium on Security and Privacy*, 104–21. https://doi.org/10.1109/SP.2015.14

Duque, J. J. (2020). State involvement in cryptocurrencies. A potential world money? *The Japanese Political Economy*, 46(1), 65–82.

Khang, A., Rana, G., Sharma, R., Goel, A. K., & Dubey, A. K. (2021). The role of artificial intelligence in blockchain applications, *Reinventing Manufacturing and Business Processes Through Artificial Intelligence*, 19–38, https://doi.org/10.1201/9781001 3145011

Khang, A., Rani, S., Chauhan, M., & Kataria, A. (2021). IoT equipped intelligent distributed framework for smart healthcare systems, *Networking and Internet Architecture*, https://arxiv.org/abs/2110.04997v2, https://doi.org/10.48550/arXiv.2110.04997

Khang, A., Rani, S., & Sivaraman, A.K. (Eds.). (2022). *AI-Centric Smart City Ecosystems: Technologies, Design and Implementation* (1st ed.). CRC Press. https://doi.org/10.1201/9781003252542

3 Framework for Modeling, Procuring, and Building Systems for Smart City Scenarios Using Blockchain Technology and IoT

Rajendra Kumar, Ram Chandra Singh, and Rohit Khokher

CONTENTS

3.1 Introduction 32
3.2 Literature Survey 32
3.3 Proposed Work 33
    3.3.1 State-of-The-Art IoT Infrastructure..33
    3.3.2 Data Analysis for Plan of Action 34
    3.3.3 Major Components of Smart City 34
        3.3.3.1 Smart Education 34
        3.3.3.2 Smart Environment 36
        3.3.3.3 Smart Roads and Platforms...37
        3.3.3.4 Smart Buildings 38
        3.3.3.5 Public Service Tracking 40
        3.3.3.6 Smart Energy..41
        3.3.3.7 Smart Agriculture...42
        3.3.3.8 Smart Banking 43
    3.3.4 IoT Devices and Support to Physically Disabled Persons..................44
3.4 Blockchain Enabled Securities 44
3.5 Blockchain and IoT Integration..44
3.6 Compendium among IoT, IoC, and IoE 46
3.7 Genesis of Smart Technologies 46
3.8 System Implementation..47
3.9 Future Challenges...47
3.10 Conclusion and Future Scope...48
References...48

DOI: 10.1201/9781003269281-3

3.1 INTRODUCTION

The term IoT was first used in 1999 for implementation of radio frequency identification system over a network. The project was proposed by MIT, Massachusetts (Xiaojie et al., 2019). Several implications and extensions have been undergone since then with the new technological advancements.

The concept of smart cities emphasizes effective use of limited resources like energy, mobility, space, etc. and troubleshooting the problems and solving them quickly applying IoT.

The IoT is a system comprising physical objects such as sensors interconnected for collecting real-time data to be transmitted for further decision making with the help of logically/physically connected machines, gadgets, and sensors. The notion of smart cities includes all things managed in smart cities way like: smart environment, smart roads and platforms, smart buildings, smart banking, smart energy, smart agriculture (Dan et al., 2015).

Over time, with development of a variety of sensors and technologies like wireless data communication, cloud computing, artificial intelligence, blockchain, Big Data, IoT, etc. are emerged and are steadily being encouraged and deployed in the secure environment (Pankaj Bhambri et al., Cloud & IoT, 2022).

IoT can play an important role in numerous areas like smart cities and is capable of helping different sectors of smart cities like smart agriculture, smart building, smart waste management (Cerchecci et al., 2018), smart healthcare, smart transport, smart environment (Kumar et al., 2015), etc.

Setting up the smart environment is important and prime but smooth function of components in secure and transparent environment is equally important. The security of a system can be ensured by installing firewall in the network and on smart devices and sensors.

To make all financial transactions transparent the use of blockchain technology would be a revolution. The computerization and tracking of all the activities of smart buildings, agriculture, roads, etc. supported by wireless sensor networks (Bannister et al., 2008), global positioning system (GPS), and control systems are to play vital role and perform expected operations in real time.

3.2 LITERATURE SURVEY

Major challenge toward implementation of smart cities is sustainable development. The architects and developers need to implement the things by considering the sustainable development goals.

Taewoo and Theresa (2011) discussed the three main dimensions (technology, citizens, and infrastructure) of smart city. The incorporation of infrastructures and smart technology as primary requirement, the establishment of smart city can be thought (Khang et al., AI-Centric Smart City Ecosystem, 2022).

Sahoo and Rath (2017) presented a detail review of major issues and applications for smart city implementation under a heterogeneous environment like cloud computing, wireless sensor networks and smart grids.

Harmon et al. (2015) discussed IoT as foundation for development of smart cities by integrating cloud-oriented architectural networks, software modules, sensors for different purposes, application interfaces as per user cantered design, data analysis in real time, etc.

Rajab and Cinkelr (2018) presented a broad review of the theories and implementation of IoT in smart cities. They described the major challenges and limitations to apply IoT technology with respect to smart city paradigms. Someone may wonder that in future a smart city may have over 50 billion objects to be connected and deployed.

Biswas and Muthukkumarasamy (2016) discussed the spread of smart technologies such as IoT, cloud services, and 24x7 network connectivity so that the smart cities can deliver innovative solutions to every citizen using direct interaction between citizens and the local government or authorities.

The concept of blockchain will make this transparent and more reliable. Despite a number of potential advantages and todays' need, digital interruption may pose many challenges with respect to data security and privacy. As a solution to this problem, Biswas and Muthukkumarasamy (2016) proposed a security framework by integrating the blockchain technology with smart devices for providing a secure communication channel in the smart city.

Rejeb et al. (2021) presented an overview of blockchain oriented applications in smart cities mainly concerned with financial transactions. The also revealed the smart city trends and suggested future research directions in this rapidly growing domain.

Naik et al. (2020) presented the contrast between current banking architecture and need of smart city banking. They also discussed vulnerabilities in the banking system to deal with frauds like the PNB scam, Kingfisher scam in India and few more. By adopting blockchain technology such fraud may be avoided in future.

Metallidou et al. (2020) considered energy performance of existing buildings by proposing an automated remote control method using cloud interface. Their method is concerned with storing energy on a cloud platform.

Nandanwar and Chauhan (2021) proposed a system to control the air and sound pollution along with monitoring and analysis based on the Internet of Things (IoT) supported by Microcontroller (like Arduino Uno R3).

Alam et al. (2021) discussed major challenges with IoT systems like security and privacy issues. They proposed blockchain as a solution to solve those issues. The integration of blockchain with IoT can provide pretty good security to the IoT based systems.

3.3 PROPOSED WORK

3.3.1 State-of-The-Art IoT Infrastructure

The required infrastructure can be accomplished using the things like a centrally managed IoT-based smart city model, a solar system capable to provide the power supply to the entire smart, a proper monitoring of the weather along with future predictions, smart agriculture, a smart waste management system, a fire and smoke detection system and a speed monitoring and traffic rules violation detection system.

A smart infrastructure is required for smart education, smart environment, smart roads and platforms, smart buildings, smart public service, smart agriculture tracking, smart energy, smart banking, etc.

FIGURE 3.1 State-of-the-art IoT enabling technologies.

The implementation of such systems depends upon exploiting the concept of IoT integration in smart city vision which includes a mixture of smart systems like noise and air pollution monitoring and controlling, fire detection systems, and smart waste bin systems with remedial solutions.

Various sensors need to be programmed associated with IoT devices to detect abnormalities by analyzing the crossed levels of climatic elements and raising alarm during danger in real time. Smart transportation for implementing the curative measures and controlling preventives will ensure accident-free movement on roads. GPS and other sensor enabled parking management for smart city residents will save time and keep vehicles safe.

State-of-the-art IoT enables the use of technologies like RFID, sensors, smart technologies, nano technologies, etc. RFID identifies and tracks the data of things (DoT). The sensors collect and process data to detect changes in physical status if any.

The nano technology works on smaller things having ability connection and interaction. This state-of-the-art IoT is presented in Figure 3.1.

3.3.2 Data Analysis for Plan of Action

The security, smart and safe working of smart city is ensured by technologies like web service management system (WSMS), fraud detection system (FDS), and total workforce management services (TWMS), safe work method statements (SWMS). WSMS is a service-oriented architecture and a standard method of sharing data among loosely coupled components of a huge system.

The FDS works with artificial intelligence and makes the network strong and safe toward intrusions. TWMS is a secured web-based system enabling for viewing, printing, and updating the specific personal employment information. SWMS includes the tasks considerably having high risk, health and safety hazards during work, control measures to minimize or remove the risks, plan of action to control the measures. The mentioned approaches enabling the proposed smart city model are shown in Figure 3.2.

3.3.3 Major Components of Smart City

Smart implementation is required everywhere whether it is education, banking and finance, energy management, agriculture, citizen management, buildings, transportation, waste, and many more as shown in Figure 3.3.

3.3.3.1 Smart Education

Smart education infrastructure and its execution is focused on use of ICT, blended mode of teaching learning, project-based learning, interactive study material facilitated by cloud services (see Figure 3.4).

Modeling, Procuring, Building Systems for Smart City Scenarios

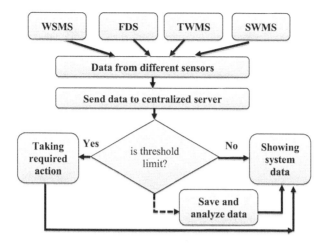

FIGURE 3.2 The flow diagram of proposed system.

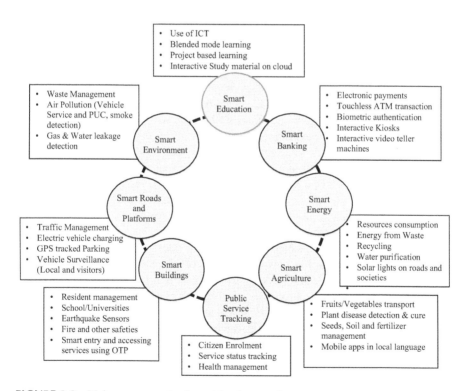

FIGURE 3.3 Major components of smart implementation.

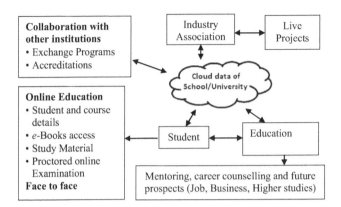

FIGURE 3.4 Smart education model.

Blockchain technology is able to keep transparency in all the modules of smart education. All financial transactions will be secured by blockchain.

3.3.3.2 Smart Environment

The smart environment in smart city is for every citizen whether front line service provider or consumer. The waste management, controlled air pollution (Vehicle Service and PUC, smoke detection), gas and water leakage detection, earthquake free buildings, and any sustainable development are taken care of in all aspects.

The regulatory bodies will provide standard operating procedures (SOPs) for all kind of sustainable developments. Smart health care will ensure a healthy life and timely action in case of illness or miss happenings.

The system for category wise smart waste bins can be implemented throughout the city. Sensors in the waste bin will decide the type of waste and accordingly the waste will be collected, or a warning alarm will be on for other category of waste. Once the bin is full, the IoT sensors will send information to a control room. This is to be monitored at central level system to send necessary instructions to collect the waste from all sub collection systems.

Apart from this, sensor enabled drones can also trace the possibility of waste on remote locations on regular intervals. On the basis of traced locations, the waste can be collected. The sensor enabled drones will also check the non-functional system and send their information to the control room for timely maintenance as shown in Figure 3.5.

The Smart waste management system will take all possible measure to detect and prevent air and noise impairments. With the help of IoT sensors, the collected information will be shared with the control room for a further plan of action.

The smoke, humidity and sound sensors will take data on regular interval to detect any panic situation. The LED bords on public places and mobile SMS alerts will make the citizens aware of unhealthy situations or danger level if they arise (see Figure 3.6).

The dotted line shows the default sharing of data for background analysis. The smart waste bins will be enabled with input voice instruction module through which the people will be able to tell the waste bin the type of waste. The waste bin module will process and understand the voice instruction and will accept or deny the type of waste.

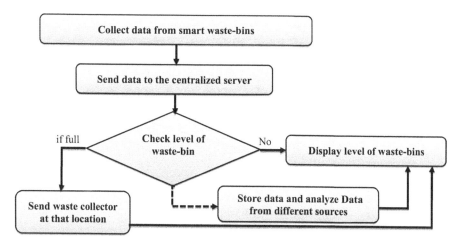

FIGURE 3.5 Proposed model for smart waste management.

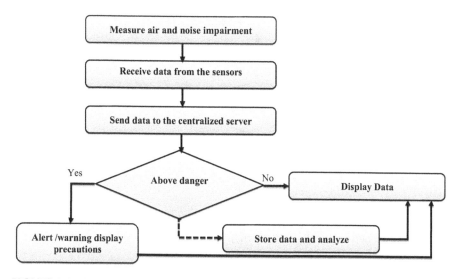

FIGURE 3.6 Proposed model for noise and air impairment detection and control.

Source: Figure is designed by Rajendra Kumar, Ram Chandra Singh, and Rohit Khokher.

3.3.3.3 Smart Roads and Platforms

The transportation department will ensure the regular circulation of do's and don'ts to the citizens of smart city. Most services will be electronic except drive test. Smart traffic management will ensure minimum causalities in roads. Electric vehicle charging stations will be smart enough to recharge using hassle-free charging.

The availability of such stations will be discoverable from mobile phones. The designated GPS tracked parking will ensure the genuine parking fee by calculating

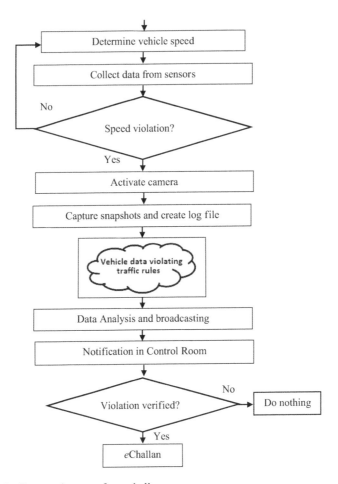

FIGURE 3.7 Proposed system for e-challan.

in and out time automatically with the help of sensors. Every vehicle will be on continuous surveillance (local and visitors) while on the road.

Entry and exit at specific points of smart city will store the log files of entry and exit details of other city's vehicles. IoT enabled vehicle speed sensors to be used to determine various parameters including speed, vehicle length, changing track, etc. Traffic violating vehicle once traced, the web cameras at designated points start snapping the images to be shared with the control room for e-challan (see Figure 3.7).

3.3.3.4 Smart Buildings

All buildings in the smart city like residential, school/universities, offices, super stores, etc. are to be smart in all aspects, whether it is a matter of installing earthquake sensors, fire and other safeties, visitor management, smart merchandise delivery using robot-controlled ducts. The smart buildings flats, shops, offices, etc. are to be assigned designated bar/QR codes for delivery of stuffs using robots. The visitor and

Modeling, Procuring, Building Systems for Smart City Scenarios

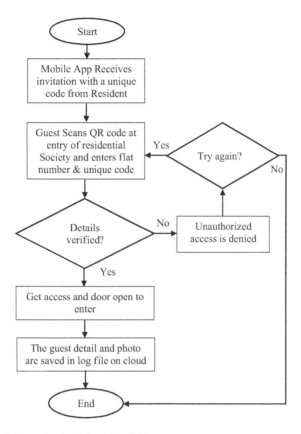

FIGURE 3.8 Automatic check-in using OTP.

Source: Figure is designed by Rajendra Kumar, Ram Chandra Singh, and Rohit Khokher.

residential management will be enabled using smart entry and accessing services using OTP to keep a record of utilization of services.

The entire city is to be equipped with sufficient access point for Wi-Fi coverage for smooth functioning of IoT devices and various sensors. The citizens will be able to connect with such access points on prepaid basis using OTP if their mobile data is not working (see Figure 3.8).

All possible and immediate efforts are needed to detect the fire in a building and timely action to stop fire. The fire to be detected by sensors and data to be shared on cloud using IoT. The control room receives an alert when new data is received on cloud.

The data is analyzed to observe the seriousness of case and local or central anti fire team is informed for necessary action. The proposed system for fire detection is shown in Figure 3.9.

The earthquake sensors in a building will continuously send the data to central servers to predict the chance and magnitude of earthquake. In case of alarming situation predicted from data, the warning messages are to be broadcasted to all the

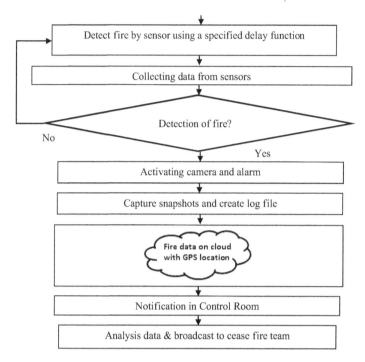

FIGURE 3.9 IoT-based fire detection system.

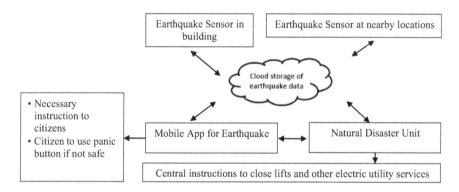

FIGURE 3.10 Proposed system for earthquake detection and plan of action.

citizens of smart city. The proposed system for earthquake detection and plan of action is presented in Figure 3.10.

3.3.3.5 Public Service Tracking
3.3.3.5.1 Citizen Management
Smart citizen management keeps a record of every citizen of the smart city. It includes the citizen enrolment module to enter newly born babies, marriage records, and death

record module to enter details of concerned persons, and a module to link all family members with each other.

The citizen enrolment module will directly fetch the newly born babies from hospitals directly and the new citizens shifting to smart city will be enrolled. This module will share the database of dead persons to disable their identities from active beneficiaries to avail various facilities.

Blockchain technology is there to transfer the bank and other funds to concerned nominees. A centralized system will include the details of all family members to track a person hierarchically.

3.3.3.5.2 Courier and other Stuff Delivery

The cables for communication networks will be passed through duct under the roads to ensure smooth maintenance. This duct will also be used for logistic purpose as well. At designated points logistics will be collected by robots by scanning bar codes. From those designated points the manual delivery of logistics will be ensured at nearby places.

All logistics will be handed over at entry of buildings. The logistics will be sent to the concerned apartment or point using conveyer belt in the duct. Every smart building will have duct for delivery of material to the concerned apartment or designated point.

As the apartment address will have a designated bar code, couriers will ensure the delivery of packages using that bar code.

3.3.3.5.3 Query Status Tracking

Smart tracking module will be enabled with blockchain technology to get the status of government schemes, employment, education, insurance, investments, etc. Health management (hospital management, medical insurance, routine check-up) will be available on mobile apps (see Figure 3.11).

The health data of every citizen will be monitored by wearable devices and in some cases swallowable pill camera capsules will record abdomen problems if any. Such smart swallowable capsules will be dissolved on expiry date. The data of capsule will be shared on IoT devices on regular intervals.

All finance (tax payment and reimbursement) managements will use blockchain technology for all the statistics of finance.

3.3.3.6 Smart Energy

For conservation of energy, the smart resources consumption plan will be implemented using IoT sensors. IoT sensors will track renewable energy waste to

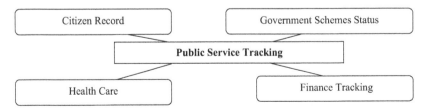

FIGURE 3.11 The major components of public service tracking.

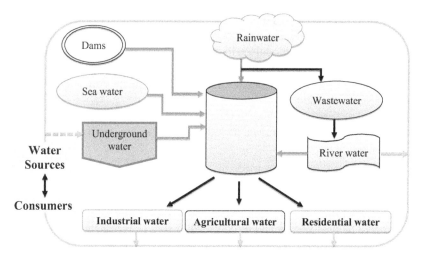

FIGURE 3.12 Water sources management.

produce energy and waste to be recycled. Used water will be converted into drinkable water. Energy sources will be tracked by IoT sensors and image processing of sources images will help the system for smooth supply chain management (see Figure 3.12).

The automatic on/off of solar streetlights will be controlled by photoelectric cells and in case of improper charging of those lights the supply of electricity will keep on the lights (*New York Times*, 2022). The roofs of every building will be covered from solar panels to fulfil the electricity need and this will also help the roof of buildings cool in summer days.

3.3.3.7 Smart Agriculture

Smart agriculture includes the use of IoT for safety and increasing productivity. It broadly includes:

- Fruits/vegetables transport – vehicles for carrying fruits and vegetables will be enabled from sensors and GPS. These will receive the information from areas with demand of fruits and vegetables. The IoT sensors will share data on mobile apps of farmers for supply of fruits, vegetables, and other materials. The previous year's production and requirements will be analyzed using machine learning algorithms to meet supply chain management of the current year (Verdouw et al., 2016; Leng et al., 2018).
- Plant disease detection will be done using deep learning algorithms (Kumar et al., 2021) and a timely cure will be ensured to retain productivity.
- Seeds, soil and fertilizer management will ensure fulfilling the needed things on time.
- Mobile apps in local language will provide farmers with the solution to their queries regarding crops, fertilizers, weather prediction and many more.

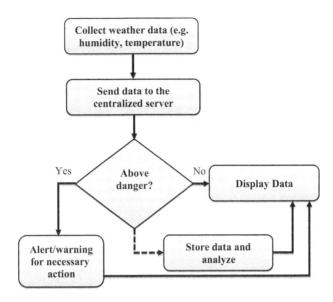

FIGURE 3.13 System to analyze weather data.

Source: Figure is designed by Rajendra Kumar, Ram Chandra Singh, and Rohit Khokher.

A dedicated outer area of smart city is for agriculture. Various sensors like temperature, humidity, fog, etc. will send data to control room and after the analysis of data the required plan of action will be executed to save the crops. The use of drones for crop quality and disease detection will be a common feature. The use of image processing and machine learning will identify the crop disease and degree of effect on crop (Foughali et al., 2018).

Remote sensing, communication technologies, IoT services, security component and management component according to their roles in agriculture will make all possible efforts to keep the crops safe and productive (Xiaojie et al., 2019). Fire detection will be observed by drones and sensors continuously. In high temperature months the crop can catch fire and farmers can go in loss.

Three major types of sensors are widely to be used in smart agricultural: physical property sensors, biosensors and micro electro-mechanical system sensors. Physical property sensors realize the signal conversion using the physical change in the sensitivity of the sensor material, which include temperature, humidity and gas (see Figure 3.13).

The biosensors (Shi et al., 2019) use the organism to transmit data according to the response of the organism to the remote location, the microbial sensors, are mainly used to detect the pesticide residue, heavy metal ion, antibiotic residue and toxic gas in the soil causing harm to crop.

3.3.3.8 Smart Banking

Smart banking will provide all citizens with electronic payments through different means. When all transactions are electronic, then cases of corruption and bribe will

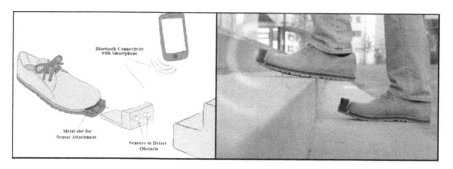

FIGURE 3.14 Sensors for blind persons.

be just negligible. The implementation of blockchain will ensure transparency in all financial transactions. For hygiene purpose, all the ATM will be updated with touchless biometric technology (Kumar et al., 2019).

In that case there will be no need to carry ATM cards for ATM transaction. The biometric identity and mobile app will enable bank customers to complete transactions (Kumar et al., 2021). In remote public places the interactive multipurpose kiosks will help users to access details of required financial schemes and stock market details.

These kiosks will also provide details and guidance regarding smart farming, fertilizer, weather, government subsidies, etc. The interactive video teller machines will help citizens in remote areas to discuss their queries with experts using video meetings.

3.3.4 IoT Devices and Support to Physically Disabled Persons

Smart shoes will help protect visually impaired people from any obstacle in their way. Sensors at the front will give them information on obstacles in audio form.

The image sensor will capture the object image to be processed by deep learning algorithms to show an alert on visually impaired people's smartphones (Khang et al., IoT & Healthcare, 2021) as Figure 3.14.

3.4 BLOCKCHAIN ENABLED SECURITIES

Blockchain is not only using bitcoin as digital currency on a mandatory basis as it is not legalized in several countries. Blockchain will enable transparency and security in all transactions using the concept of distributed ledger of all finances.

The distributed ledger in blockchain provides interoperability, trust and transparency. In healthcare and other government schemes, a person can give the proof for the access of his/her information to maintain transparency.

3.5 BLOCKCHAIN AND IOT INTEGRATION

The smart city will use smart systems for different purposes (see Figure 3.15). All such systems (machines, devices, home appliances, and other embedded systems)

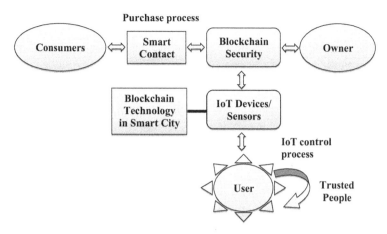

FIGURE 3.15 Blockchain and IoT integration.

Source: Figure is designed by Rajendra Kumar, Ram Chandra Singh, and Rohit Khokher.

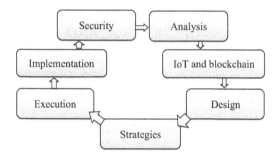

FIGURE 3.16 Life cycle of relationship in blockchain and IoT.

will be strengthened to deal with real-time sharing of information to control room and timely action as instructed by control room authorities.

Sufficient distributed sensors will share information to centralized and distributed servers for further processing. The security of all resources will be taken care by blockchain technology (Khang et al., AI & Blockchain, 2021).

IoT will play an important role in managing all resources and prevent any kind of disruption in all commercial and business practices. The practical model implantation is based on the blockchain life cycle to ensure blockchain processes, business-process-model (BPM) related information technology, process execution, etc. However, blockchain technology might affect the existing e-governance due to identification and distribution of new blocks over the network.

The blockchain lifecycle includes seven different processes: analysis, IoT and blockchain, design, strategy, execute, implement, security and analysis. These seven cyclic components of blockchain life cycle process in the proposed smart city model are shown in Figure 3.16.

The following are the main proposed activities of the blockchain lifecycle:

1. The analysis part contains acquisition insights of issues related to business processes.
2. IoT security enforces blockchain rules for business-process-model based information technology.
3. Design deals with identification and distribution of the blocks over the shared network.
4. Strategies derived from standard operating procedures.
5. Execution deals with processing of individual cases related to information technology.
6. Implementation deals with procedure of transforming the model into the system or process.
7. Monitoring security of processes implemented.

Using blockchain, there will be transparency in documents, greater scalability in number of participants involved in different transactions, transparent audits, increased innovations, condensability, integrity, etc. (El-Din et al., 2018).

3.6 COMPENDIUM AMONG IOT, IOC, AND IOE

The integration of IoT and internet of computing (IoC) with internet of everything (IoE) will enable the use of big data, cloud computing, social computing and mobility as Figure 3.17.

3.7 GENESIS OF SMART TECHNOLOGIES

The model proposed in this chapter is an integration of systems constituted to monitor and control the dangers from air, noise, temperature, weather, fire due to short circuits and other means, etc.

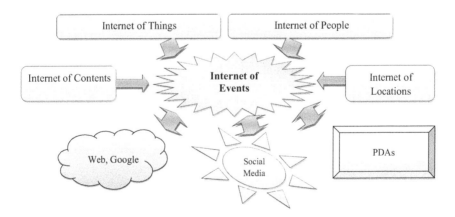

FIGURE 3.17 The integration of various things.

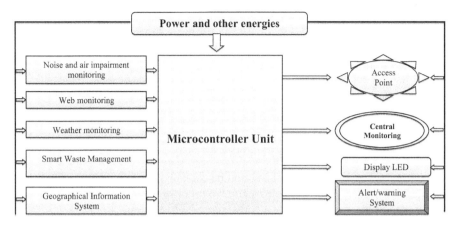

FIGURE 3.18 Proposed framework for smart cities.

It also encompasses smart waste bin enabled with sensors and smart delivery, tracking and transportation using ducts below the road level and smart building supported by programmed robots. Smart energy will minimize the wastage and will optimize the utilization. Energy utilization will be accessible to all citizens of smart city through blockchain technology.

The key to the success of the model is to do smart everything from the beginning. Various microcontrollers equipped with IoT devices will monitor the activities and wherever required the improved solutions will be explored and applied.

3.8 SYSTEM IMPLEMENTATION

Figure 3.18 shows the implementation of the proposed model of smart city consisting of sensors for different purposes like temperature, humidity, smoke, sound, motion, fire, traffic rules violation, webcam, smart water bin, energy conservation, healthcare, crop inspection, stuff tracking, etc.

All those sensors are connected with microcontroller unit (MCU) to receive the data and timely required actions. The proposed model collects the data from different sensors and sends to the server to process centrally. All observations and actions taken are to be stored on the cloud to analyze for better practices in future.

3.9 FUTURE CHALLENGES

The major challenges are related to troubleshooting of IoT devices, their connectivity, timely replacement and/or upgradation. Continuous efforts are required for scalability, security, complexity, and speed cost and domain dependence.

Continuously growing size of blockchain network may cause security attacks. Scalable blockchains may deal with the big size of blockchain network growing continuously. The smart city planner will require implementation of distributed management systems to detect and correct issues zone wise. The complexity of blockchain has transformed more typical cryptography. Confidentiality is another factor to be taken care of in data protection.

For perfect implementation of the smart city, the thing most required is mindset. Everyone will have to trust the system and be honest by showing readiness, enthusiasm from day one. Some possibility of human error is always there. This becomes more complicated to troubleshoot with blockchain as it mixes software skills to deal with economics. Trained IT staff is required to solve daily issues.

3.10 CONCLUSION AND FUTURE SCOPE

This chapter focused on the use of IoT and blockchain in day-to-day transparent and recorded activities of citizens of smart city. Smart citizen management will track the unauthorized activities at all the levels (Kumar, Singh, & Kant, 2021).

The smart city is a beginning and always there will be scope for further improvement in using sensors, algorithms implementing subsystems and systems, security solutions, interactive kiosks, communication media, embedded systems, deep learning-based healthcare systems and agriculture, waste conversion into energy, wearable healthcare devices, smart transport making enjoyable trip of disabled persons, etc.

The smart city is a practice of resource conservation and perfect utilization of resources and services by saving time and efforts, making human life enjoyable, physically, and financially secured, detecting unpleasant acts before they happen using sensors and a timely plan of action to prevent them.

The concept of smart city is for all, whether it is a farmer doing farming smartly, front line worker serving in healthy way, all concerned paying taxes to strengthen the government economy, benefits of government schemes to all needy persons in a transparent way.

Future work can be focused to deal with solutions to fulfil the challenges in troubleshooting of subsystems, sensors, and physical conditions.

REFERENCES

Alam, S. R., Jain, S., & Doriya, R., Security threats and solutions to IoT using blockchain: A review. In *Proceedings of the 5th International Conference on Intelligent Computing and Control Systems (ICICCS)*, 2021, pp. 268–73.

Bannister, K., Giorgetti, G., & Gupta, S. K., Wireless sensor networking for hot applications: Effects of temperature on signal strength, data collection and localization. In *Proceedings of the 5th Workshop on Embedded Networked Sensors, Charlottesville, VA, USA*, June 2–3, 2008, pp. 1–5.

Bhambri, Khang, A., Sita Rani, & Gaurav Gupta, *Cloud and Fog Computing Platforms for Internet of Things*. Chapman & Hall. ISBN: 978-1-032-101507, 2022, https://doi.org/10.1201/9781032101507.

Biswas, K., & Muthukkumarasamy, V., Securing smart cities using blockchain technology. In *Proceedings of the IEEE 18th International Conference on High Performance Computing and Communications; IEEE 14th International Conference on Smart City; IEEE 2nd International Conference on Data Science and Systems (HPCC/SmartCity/DSS)*, 2016, pp. 1392–1393.

Cerchecci, M., Luti, F., Mecocci, A., Parrino, S., Peruzzi, G., & Pozzebon, A. A low power IoT sensor node architecture for waste management within smart cities context, *Sensors*, 2018, 18, 1282.

El-Din, D. M., Hamed, M., & Khalifa, N. E., A blockchain technology evolution between business process management (BPM) and Internet-of-Things (IoT), *International Journal of Advanced Computer Science and Applications*, 2018, 9(8), pp. 442–50.

Foughali, K., Fathallah, K.M., & Frihida, A. Using cloud IOT for disease prevention in precision agriculture. *Procedia Computer Science*, 2018, 130, pp. 575–82.

Harmon, R. R., Castro-Leon, E. G., & Bhide, S., Smart cities and the Internet of Things. In *Proceedings of the Portland International Conference on Management of Engineering and Technology (PICMET)*, 2015, pp. 485–94.

Khang, A., Geeta Rana, Ravindra Sharma, Alok Kumar Goel, & Ashok Kumar Dubey, The role of artificial intelligence in blockchain applications, *Reinventing Manufacturing and Business Processes Through Artificial Intelligence*, pp. 19–38, 2021, https://doi.org/10.1201/9781003145011

Khang, A., Rani, S., & Sivaraman, A.K. (Eds.). (2022). *AI-Centric Smart City Ecosystems: Technologies, Design and Implementation* (1st ed.). CRC Press. https://doi.org/10.1201/9781003252542

Khang, A., Sita Rani, Meetali Chauhan, & Aman Kataria, IoT equipped intelligent distributed framework for smart healthcare systems, *Networking and Internet Architecture*, 2021, https://doi.org/10.48550/arXiv.2110.04997

Kumar, R., Singh, R. C., & Kant, S., Dorsal hand vein-biometric recognition using convolution neural network, *Advances in Intelligent Systems and Computing*, 2021, pp. 1087–1107.

Kumar, R., Singh, R. C., & Khokher, R., Dorsal hand vein based touchless system for ATM cash withdrawal, App. No. 202111017902, published on April 23, 2021.

Kumar, R., Singh, R. C., & Sahoo, A. K., SIFT based dorsal vein recognition system for cashless treatment through medical insurance, *International Journal of Innovative Technology and Exploring Engineering*, 8(10S), 2019, pp. 444–51.

Kumar, V., & Ramya, R., The real time monitoring of water quality in IoT environment. In *Proceedings of the International Conference on Innovations in Information, Embedded and Communication Systems, Coimbatore, India*, 2015, pp. 1–5.

Metallidou, C. K., Psannis, K. E., & Egyptiadou, E. A., Energy efficiency in smart buildings: IoT approaches, *IEEE Access*, vol. 8, 2020, pp. 63679–99.

Naik, V., Pejawar, R., Singh, R., Aher, A., & Kanchan, S., Expeditious banking using blockchain technology. In *Proceedings of the International Conference on Computational Intelligence for Smart Power System and Sustainable Energy (CISPSSE)*, 2020, pp. 1–6.

Nandanwar, H., & Chauhan, A., IOT based smart environment monitoring systems: A key to smart and clean urban living spaces. In *Proceedings of Asian Conference on Innovation in Technology (ASIANCON)*, 2021, pp. 1–9.

Rajab, H., & Cinkelr, M.T., IoT based smart cities, In *Proceedings of International Symposium on Networks, Computers and Communications (ISNCC)*, 2018, pp. 1–4.

Rejeb, A., Rejeb, K., & Simske, S. J., Blockchain technology in the smart city: A bibliometric review, *Qual Quant*, 2021. https://doi.org/10.1007/s11135-021-01251-2

Sahoo, J., & Rath, M., Study and analysis of smart applications in smart city context. In *Proceedings of the International Conference on Information Technology (ICIT)*, 2017, pp. 225–8.

Shi, X., An, X., Zhao, Q., Liu, H., Xia, L., Sun, X., & Guo, Y., State-of-the-art Internet of Things in protected agriculture, *Sensors*, 19(8), 2019, 1833.

Taewoo, N., & Theresa, A. P., Conceptualizing smart city with dimensions of technology, people, and institutions. In *Proceedings of the 12th Annual International Conference on Digital Government Research, College Park, MD, USA*, 2011.

The New York Times, 2022, Bitcoin Uses More Electricity Than Many Countries. How Is That Possible? https://www.nytimes.com/interactive/2021/09/03/climate/bitcoin-carbon-footprint-electricity.html (accessed Feb. 08, 2022)

Verdouw, C. N., Wolfert, J., Beulens, A. J. M., & Rialland, A., Virtualization of food supply chains with the Internet of Things. *Journal of Food Engineering*, 2016, 176, 128–36.

Xiaojie, S., Xingshuang, A., Qingxue, Z., Huimin, L., Lianming, X., Xia, S., & Yemin, G., State-of-the-art Internet of Things in protected agriculture, *Sensors*, 2019, 19(8), 1833, pp. 1–24.

4 Development of a Framework Model Using Blockchain to Secure Cryptocurrency Investment

K. Mohamed Jasim, Manivannan Babu, and Chinnadurai Kathiravan

CONTENTS

4.1 Introduction 51
4.2 Blockchain Technology 52
4.3 Need for Blockchain Technology in Cryptocurrency 53
4.4 Model to Secure Cryptocurrency Investment 53
4.5 Transaction Unified as Block 55
4.6 Block as Network 55
4.7 Network Approval for Transaction 55
4.8 End–End Encryptions 56
    4.8.1 Decentralization 56
    4.8.2 Trust 56
    4.8.3 Security 56
4.9 Discussion 57
4.10 Conclusion 57
References 58

4.1 INTRODUCTION

Success of bitcoin and another cryptocurrency like Ethereum, Litecoin etc. attracts the attention of technocrats in blockchain technology. People have positive and negative opinions on it. For instance, Tapscott and Tapscott fear its negative effects on the financial system (Tapscott & Tapscott, 2016).

Despite significant impact on the economy, it was late to get the attention of academic research on cryptocurrency and blockchain technology (Karame, et al., 2105; Cheah & Fry, 2016; Polasik, Piotrowska, Wisniewski, Kotkowski, & Lightfoot; Bouiyour & Selmi, 2015; Kristoufek, 2013; Garcia, Tessone, Mavrodiev, & Perony, 2014).

DOI: 10.1201/9781003269281-4

Price fluctuation in cryptocurrency is the big issue in adopting it as a medium of transaction. As far as investment decision making is concerned pricing the cryptocurrency and understanding the reason of fluctuation in price are still an issue of big debates. This is the reason some researchers consider it as a speculative bubble and not a currency for transaction (Kristoufek, 2013; Garcia, Tessone, Mavrodiev, & Perony, 2014; Bohme, Christin, Edelman, & Moore, 2015). Determination and prediction of the value of cryptocurrency are still a mystery for investment decision making.

The current study attempts to determine the value of cryptocurrency, particularly bitcoin, based on the literature review and technological aspect and acceptance of monetary economic theories.

In this study cryptocurrency has been evaluated as technological invention and artifact and as an instrument of economic value and means of value transfer. This systematic discussion on the multiple aspect of the cryptocurrency and blockchain technology will help in synthesizing the existing evidence and will become the base for future discussion and model extension research. Considering this as the research gap, a blockchain model has been proposed for the cryptocurrency industry (Douceur, 2002).

4.2 BLOCKCHAIN TECHNOLOGY

Blockchain technology provides a fair, democratic and fully publicly maintained system of data storage and authorization. It is highly secure because it does not have one point of failure (Armbrust et al., 2010).

Hackers need only one system to hack in case of centralized system of data storage and authorization (Pankaj Bhambri et al., Cloud & IoT, 2022). However, in the case of blockchain technology hackers need to hack thousands of systems or nodes in case of millions of nodes in the network. Apart from the data structure this technology is also known for the digital consensus architectures through proof of work (PoW) (Mattila et al., 2016).

PoW is an algorithm or domain of application built on top of such architectures. The digital ledger is accessible to everyone and therefore a proof of work is required. When a transaction take place it comes in a pool of all the transactions and it is in an unconfirmed state. To confirm the transactions and to write it in the block in a blockchain ledger a node individually or several nodes collectively must solve the puzzle to write the block. This requires time and energy that counts against the cost. The nodes of those who are involved in the PoW process are called miners and they get the reward if they win the PoW (Khang et al., AI & Blockchain, 2021; Khang et al., The Data-Driven Blockchain Ecosystem, 2022).

The crash of the Central Bank of different government during 2008 was the main reason for the popularity of cryptocurrencies (Weber, 2014). The transaction cost of cryptocurrencies is also cheaper than debit and credit cards (Angel & McCabe, 2015).

The academician participation in the research related to bitcoin started late and mostly surrounded by the use of bitcoin (Böhme, Brenner, Moore, & Smith, 2014;

Sadeghi, 2013) and legal status of the bitcoin or cryptocurrency (Grinberg, 2012; Plasaras, 2013).

Now academic interest has been expended and discussion has started on the different aspects of the bitcoin including financial aspects and its impact on the economy. Previous literature discussed several exciting questions: for instance, the exact nature of cryptocurrency or digital currency, its long-term sustainability and legal and ethical aspects (Bearman, 2015). This chapter also contributes to these discussions and adds to a burgeoning tectonophysics literature on bitcoin and cryptocurrencies.

4.3 NEED FOR BLOCKCHAIN TECHNOLOGY IN CRYPTOCURRENCY

The idea of virtual currency is not a new concept: several time innovators attempted to start virtual money, some examples being web-based money, B-money (Dai, 1998), Bit gold, Hashcash etc. (Khang et al., The Data-Driven Blockchain Ecosystem, 2022).

Chaum in his research paper presented an idea of public key cryptography that can hide the content as well as the sender information through unsecured telecommunication system. In this system there is no need for trusted third-party involvement (Chaum, 1981).

Law et al. present the cryptography technique used to manage electronic cash transactions where a private key is used to sign the transaction and public key used to verify the same. But in his approach, he suggested the central authority to remain in the loop and authenticate the transaction (Law, Sabett, & Solinas, 1996).

First time proof of work was suggested by Dwork and Naor in 1992. He wanted to use this system against junk mail to allow the receiver to verify the competition. This might be the first idea of providing proof of work in electronic data transfer (Dwork & Naor, 1992).

4.4 MODEL TO SECURE CRYPTOCURRENCY INVESTMENT

Bitcoin has power by design to make it a real currency which takes years to get recognition; furthermore, the total number of currencies is limited to 21 million (King, 2013). This facet of bitcoin is important for its value.

The number of currencies is fixed to eliminate inflation due to extra currency in the market. The national government restriction and changes in the financial policy do not result in inflation in cryptocurrency (Magro, 2016).

Due to having restraint from inflation, bitcoin or other cryptocurrency may be the best place of investment but like other goods and commodities the value of cryptocurrency also fluctuates due to several other reasons. Investors make the bitcoin as the highest value money in the world in 2015 using the Dollar Index (Desjardins, 2016).

The layout of the bitcoin is not free from error and weakness. It is based on the technology of blockchain and public ledger that means everyone can see and track the transactions. However, the owner may not be identified by its name but through its public key.

If anyone want to track the transaction through any public key, he/she may track all the transaction. The open ledger database (blockchain) is shared with every user of the cryptocurrency, which makes it vulnerable to attack (King, 2013) but at present due to a large number of systems participating in the blockchain pool it is almost impossible to change the data.

Bitcoin during its journey has gone through several tests; essentially these are attacks such as DDoS attacks (Hileman, 2016). Some of them are relatively minor to prove their point about bitcoin design.

The participant proof that they can jointly bring the bitcoin network down is not a good design for future use. However, these are integral with the bitcoin and cannot be changed. Another demerit of the bitcoin is its uses for unauthorized business and transactions (Khang et al., AI-Centric Smart City Ecosystem, 2022).

Bitcoin has a negative image among government bodies due to its inability to restrict unwanted transactions, as shown in Figure 4.1.

One prospective risk to cryptocurrencies originates from its speculative nature. A large number of people invest in one or other cryptocurrency to get the profit due to increase in value. This results in extreme flection in the price of cryptocurrency and market crash (Bitshares, 2022).

The selection of design of the internet exchanges where cryptocurrencies are traded might also contribute to these impacts if aspects of readily available performance, graphical user interfaces (GUI), or application programs user interfaces (API) promote cumulative exhilaration. Markets are human artifacts, not natural phenomena, and also as a result a target of design (Lampinen and Brown, 2017).

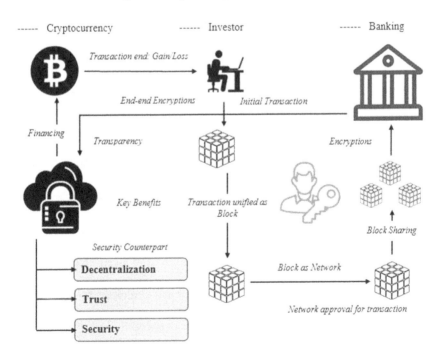

FIGURE 4.1 Blockchain model.

In the present research we make every effort to much better recognize the variables that contribute to collective excitement in cryptocurrencies, and also exactly how the design of cryptocurrency market systems as well as user interfaces might influence these procedures (Billah, 2015).

Central to these goals is comprehending why people at a certain time decide to buy a particular modern technology, product, or idea. If the assets are new, or information concerning them has just been launched, financial investment could be a rational feedback to today's state of information (Barber and Odean,)2008.

Various other elements could include authorities backing the investment, or huge players making visibly big bets on it (Bikhchandani and Sharma, 2000). One more hypothesized resource of cumulative positive outlook is peer impact amongst little private traders (Spyrou, 2013; Hirshleifer and Teoh, 2003).

4.5 TRANSACTION UNIFIED AS BLOCK

A user of bitcoin has to enter the amount or value and receiver address to generate the transaction. The transaction is then signed by the user using its private key. The other neighboring nodes check the validity of the transactions and forward them to the other nodes if the transactions are valid (Antonopoulos, 2014).

MultiChain includes added metadata to the transaction, defining asset name and transaction type (providing or investing properties, granting permissions etc.). The deal is constructed, authorized and also transmitted to the network. In Chain Core, transactions create new assets or spend existing one (Moskowitz and Swanson, 2015).

4.6 BLOCK AS NETWORK

The user of cryptocurrency owns a pair of keys known as private and public key. The private key is like a password and is used to validate the transaction. Once the transaction is validated it is broadcast through the whole network where it is checked for validity and the transaction is written in the blockchain permanently using some validation algorithm. A blockchain is the open ledger which holds all the record from the first transaction (Khang et al., AI & Blockchain, 2021).

The first block in the blockchain is called the genesis block, which has no parent block. Therefore it continues to grow as time is spent and transactions increase. Figure 4.1 illustrates an example of a blockchain. Every block has a hash field and contains the hash field of the parent block. It is worth noting that Ethereum includes the address of the branches of the parent block (Buterin, 2014).

4.7 NETWORK APPROVAL FOR TRANSACTION

Blockchain technology is considered a trusted ledger due to two features: one is accuracy and the other is immutability. Due to these features a transaction does not require any trusted intermediary to establish trust among the persons involved in the transactions even when they do not know each other and do not want to reveal their identity.

The *Economist* magazine calls the blockchain the "Faith Machine" and Finck (2017) describes it as "trustless faith, which makes it possible to trust the results of a system without relying on any actor within it." It still appears that when blockchains have to link with the genuine world to obtain details concerning possessions, a relied on authority might be needed to make that link.

4.8 END–END ENCRYPTIONS

The validation process of the blockchain process may be attacked by the more than 50 percent colluding miners. In particular, Eyal and Sirer (2014) showed that the network is vulnerable even if only a small portion of the hashing power is used to cheat. In self-seeking mining technique, selfish miners maintain their extracted blocks without broadcasting and the exclusive branch would certainly be exposed to the public only if some requirements are satisfied.

As the personal branch is longer than the existing public chain, it would be confessed by all miners. Prior to the exclusive blockchain publishment, honest miners are wasting their sources on a pointless branch while self-seeking miners are mining their exclusive chain without rivals. Selfish miners often tend to get more income. Based on self-centered mining, lots of other attacks have actually been suggested to reveal that blockchain is not so safe (Nayak et al., 2016).

4.8.1 Decentralization

The blockchain techniques potentially allow individuals and groups to revamp their interactions in politics, company and society at large, with an extraordinary process of disintermediation on big range, based on automated and trustless deals. This procedure may swiftly change even the tenets that underpin existing political systems and governance models, calling into question the traditional role of State and centralized institutions (European Central Bank, 2012). Many blockchain advocates declare that civil society might organize itself and secure its own interests extra efficiently, by replacing the conventional features of State with blockchain-based services and decentralized, open source platforms (e.g. bitcoin, Ethereum) (Paquet & Wilson 2015).

4.8.2 Trust

Trust is a basic factor in any interaction that occurs between people (Gambetta, 1988) and widely recognized as a multi-faceted idea with cognitive, psychological, and behavioral dimensions (Lewis & Weigert, 1985). The concept of confidence is securely anchored in several academic disciplines: psychology, ideology, and economics (Baier, 1986; Good, 1988; Sako, 1992) bring about a selection of interpretations.

4.8.3 Security

In spite of the open-source nature of protocols and the much-vaunted egalitarianism of peer-to-peer networks, a huge adoption of blockchain solutions without public organizations to coordinate their activity would most likely wind up developing new oligarchies and a strong polarization in society (Khang et al., IoT & Healthcare, 2021).

4.9 DISCUSSION

In blockchain, how to reach consensus among the unreliable nodes is a transformation of the Byzantine Generals (BG) Problem, which was raised in (Lamport et al., 1982). In the BG problem, a group of generals who command a part of a Byzantine military circle the city. Some generals like to strike while various other generals like to pull away. The attack would fall short if just component of the generals strike the city.

Hence, they have to reach a contract to attack or retreat. Just how to get to an agreement in dispersed environment is a difficulty. It is additionally an obstacle for blockchain as the blockchain network is distributed. In blockchain, there is no central node that makes sure that journals on distributed nodes are just the same (Ciaian et al., 2016).

Some protocols are needed to make certain ledgers in various nodes consistent. We next use existing several typical strategies to get to an agreement in blockchain. PBFT (Practical Byzantine Mistake Resistance) is a replication algorithm to enable tolerance of Byzantine mistakes (Miguel and Barbara, 1999).

Hyperledger Fabric (Hyperledger task, 2015) utilizes the PBFT as its agreement formula because PBFT can take care of up to 1/3 malicious Byzantine replicas. A brand-new block is figured out in a round. In each round, a key is chosen according to certain guidelines (Nathan et al., 2018).

4.10 CONCLUSION

Tapscott and Tapscott (2016) compare blockchain technology in its value to the development of the net. The strong public interest develops from the one-of-a-kind features of blockchains. As an open and dispersed ledger that can record transactions successfully, completely and verifiably, blockchain innovation has the possibility to open new company designs and make standard ones obsolete.

Middlemen such as lawyers, banks, or notaries may not be required in the future anymore. The blockchain was first called a part of the bitcoin procedure by Nakamoto (2008). Bitcoin needed a method to capture and confirm operations and the blockchain was the solution (Fanning & Centers, 2016).

In the years since, research study on blockchain applications generally concentrated on bitcoin (Yli-Huumo et al., 2016). It has explored technological aspects, such as different blockchain types, agreement systems, cryptocurrencies, and governance systems (De Kruijff & Weigand, 2017), and has barely explored trust relevant issues. Nonetheless, reliability appears to be a vital aspect in the context of possible blockchain-driven improvements.

This recommends further discovering the effect that blockchain technology can carry the role of rely on transactions. Here, in this qualitative chapter we aim to make a payment. We lay out the academic version for blockchain technology which is extremely suitable for cryptocurrency sectors and focus on terms and theoretical trust related concepts for exploring how the blockchain influences the function of count on exchanges along 4 attributes of reliability: trustee, activity and trustor partnership, susceptibility, and the subjective issue (Wang & Emurian, 2005).

REFERENCES

Angel, J. J., & McCabe, D. (2015). The ethics of payments: paper, plastic or Bitcoin? *Journal of Business Ethics* 132, 603–11.

Antonopoulos, A.M. (2014). *Mastering Bitcoin: Unlocking Digital Crypto-Currencies*. O'Reilly Media, Inc., 1st edn.

Armbrust, M., Fox, A., Griffith, R., Joseph, A.D., Katz, R., Konwinski, A., et al. (2010). A view of cloud computing. *Commun ACM* 53(4), 50–8.

Baier, A. (1986). Trust and antitrust. *Ethics* 96(2), 231–60.

Barber, Brad M., & Odean, T. (2008). All that glitters: the effect of attention and news on the buying behavior of individual and institutional investors. *Review of Financial Studies* 21(2), 785–818.

Bearman, J. (2015, May). The Untold Story of Silk Road, Pt. 1. Retrieved from Wired.com Website: www.wired.com/2015/04/silk-road-1/

Bhambri, P., Rani, R., Gupta, G., & Khang, A. (2022). *Cloud and Fog Computing Platforms for Internet of Things*. Chapman & Hall. ISBN: 9781032101507, doi: 10.1201

Bikhchandani, Sushil, & Sunil Sharma (2000). Herd behavior in financial markets. *IMF Staff Papers*, 279–310.

Billah, S. (2015). One weird trick to stop selfish miners: Fresh bitcoins, a solution for the honest miner, 2015.

Bitshares – your share in the decentralized exchange. (2022 [Online]). Available: https://bitshares.org/

Böhme, R., Brenner, A., Moore, T., & Smith, M. (Eds.) (2014). *Financial Cryptography and Data Security*: FC 2014 Workshops, BITCOIN and WAHC 2014, Christ Church, Barbados, March 7, 2014, Revised Selected Papers.

Bohme, R., Christin, N., Edelman, B.G., & Moore, T. (2015). Bitcoin: Economics, technology, and governance. *Journal of Economic Perspectives* 29(2), 213–38.

Bohme, R., Christin, N., Edelman, B., & Moore, T. (2015). Bitcoin: Economics, technology,and governance. *Journal of Economic Perspectives* 29(2), 213–38.

Buterin, V. (2014). A next-generation smart contract and decentralized application platform, white paper.

Chaum, D.L. (1981). Untraceable electronic main, return addresses, and digital pseudonyms. *Communications of the ACM* 24(2), February, 84–8.

Cheah, E.T., & Fry, J. (2015). Speculative bubbles in Bitcoin markets? An empirical investigation into the fundamental value of Bitcoin. *Economics Letters* 130, 32–6.

Chuen, Lee Kuo. Ed. (2015). *Handbook of Digital Currency*, 1st ed. Elsevier. [Online]. Available: http://EconPapers.repec.org/RePEc: eee: monogr: 9780128021170

Ciaian, P., Rajcaniova, M., & Kancs, D. (2016). The economics of Bitcoin price formation. *Applied Economics* 48(19), 1799–1815.

Dai, W. (1998). B-money. 1998 Available at: www:weidai:com/bmoney.

Desjardins, J. (2016, January 5). It's Official: Bitcoin was the Top Performing Currency of 2015. Retrieved from The Money Project Website: http://money.visualcapitalist.com/its-official-bitcoin-was-thetopperforming-currency-of-2015/

Douceur (2002). The Sybil attack. In *Proceedings of IPTPS '01 Revised Papers from the First International Workshop on Peer-to-Peer Systems*, pp. 251–60, March.

Dwork, & M. Naor (1992). Pricing via processing or combatting junk mail. In *12th Annual International Cryptology Conference*, pp. 139–47, 1992.

European Central Bank (2012). Virtual currency schemes. Preprint www.ecb.europa.eu/pub/pdf/other/virtualcurrencyschemes201210en.pdf

Eyal, & E. G. Sirer (2014). Majority is not enough: Bitcoin mining is vulnerable. In *Proceedings of International Conference on Financial Cryptography and Data Security, Berlin, Heidelberg*, pp. 436–54.

Fanning, K., & D. Centers (2016). Blockchain and its coming impact on financial services. *Journal of Corporate Accounting & Finance* 27(5), 53–7.

Finck, Michèle (2017). Blockchain Regulation. Max Planck Institute for Innovation and Competition Research Paper (17–13).

Gambetta, D. (1988). Can we trust trust? In: *Trust – Making and Breaking Cooperative Relations*. Ed. by D. Gambetta. New York: Basil Blackwell, 213–37.

Garcia, D., Tessone, C. J., Mavrodiev, P., & Perony, N. (2014). The digital traces of bubbles: Feedback cycles between socio-economic signals in the Bitcoin economy. *Journal of the Royal Society Interface* 11(99). https://doi.org/10.1098/rsif.2014.0623

Good, D. (1988). Individuals, interpersonal relations, and trust. In: *Trust – Making and Breaking Cooperative Relations*. Ed. by D. Gambetta. New York: Basil Blackwell, 31–48.

Hileman, G. (2016, January 28). State of Bitcoin and Blockchain 2016: Blockchain Hits Critical Mass. Retrieved from Coindesk Website: www.coindesk.com/state-of-bitcoin-blockchain-2016/

Hirshleifer, David, & Siew Hong Teoh (2003). Herd behaviour and cascading in capital markets: A review and synthesis. *European Financial Management* 9(1), 25–66.

Hyperledger project (2015 [Online]). Available: www.hyperledger.org/

Karame, G., Roeschlin, M., Gervais, A., Caplun, S., Androulaki, E., & Caplun, S. (2015). Misbehavior in bitcoin: A study of double-spending and accountability, *ACM Transactions on Information and System Security* 18(1), 2.

Khang, A., Geeta Rana, Ravindra Sharma, Alok Kumar Goel, & Ashok Kumar Dubey (2021). The role of artificial intelligence in blockchain applications, *Reinventing Manufacturing and Business Processes Through Artificial Intelligence*, doi: 10.1201/9781003145011

Khang, A., Sita Rani, & Arun Kumar Sivaraman (2022). *AI-Centric Smart City Ecosystem: Technologies, Design and Implementation*. HB. ISBN: 978-1-032-17079-4 ** EB.ISBN: 978-1-003-25254-2, doi: 10.1201/9781003252542

King, R. S. (2013, December 17). By reading this article, you're mining bitcoins. Retrieved from Quartz.com Website: http://qz.com/154877/byreading-this-page-you-are-mining-bitcoins/

Kristoufek, L. (2013). Bitcoin meets Google trends and Wikipedia: Quantifying the relationship between phenomena of the Internet era. *Scientific Reports*. doi:10.1038

Kristoufek, L. (2013). Bitcoin meets Google trends and Wikipedia: Quantifying the relationship between phenomena of the Internet era. *Scientific Reports* 3, 3415.

Lampinen, Airi, & Barry Brown (2017). Market design for HCI: Successes and failures of peer-to-peer exchange platforms. In *Proceedings of the ACM Conference on Human Factors in Computing Systems (CHI). ACM*, 4331–43.

Lamport, R. Shostak, & M. Pease (1982). The Byzantine generals problem. *ACM Transactions on Programming Languages and Systems (TOPLAS)* 4(3), 382–401.

Law, S. Sabett, & J. Solinas (1996). How to make a mint: the cryptography of anonymous electronic cash. *American University Law Review* 46(4), 1131–62.

Lewis, J., & A. Weigert (1985). Trust as a social reality. *Social Forces* 63(4), 967–85.

Magro, P. (2016, July 16). What Greece can learn from bitcoin adoption in Latin America. Retrieved July 2016, from International Business Times Website: www.ibtimes.co.uk/what-greece-can-learn-bitcoinadoptionlatin-america-1511183

Mattila, J. (2016). The blockchain phenomenon-the disruptive potential of distributed consensus architectures. www.brie.berkeley.edu/wpcontent/uploads/2015/02/Juri-Mattila-.pdf

Mattila, J., Seppälä, T., Naucler, C., Stahl, R., Tikkanen, M., Bådenlid, A., et al. (2016). Industrial blockchain platforms: An exercise in use case development in the energy industry. www.etla.fi/julkaisut/industrial-blockchainplatforms-an-exercise-inuse-case-development-in-the-energy-industry

Miguel & L. Barbara (1999). Practical Byzantine fault tolerance. In *Proceedings of the Third Symposium on Operating Systems Design and Implementation*, vol. 99, New Orleans, pp. 173–86.

Nakamoto, S. (2008). Bitcoin: A peer-to-peer electronic cash system. [Online]. Available: https://bitcoin.org/bitcoin.pdf

Nathan, S., Thakkar, P. & Vishwanathan, B. (2018). Performance benchmarking and optimizing hyperledger fabric blockchain platform, in Proc. 2018 IEEE 26th International Symposium on Modeling, Analysis, and Simulation of Computer and Telecommunication Systems (MASCOTS), pp. 264–76.

Nayak, S., Kumar, A. Miller, & E. Shi (2016). Stubborn mining: Generalizing selfish mining and combining with an eclipse attack. In *Proceedings of 2016 IEEE European Symposium on Security and Privacy (EuroS&P), Saarbrucken, Germany*, pp. 305–20.

Paquet, G., & Wilson, C. (2015). Governance Failure and Antigovernment Phenomena. CoG Working Paper, June 2015. Retrieved from www.patheory.net/conference2015/papers/antigovernment-asgovernance-failure-ver-12.pdf.

Plasaras, N. (2013). Regulating digital currencies: Bringing bitcoin within the reach of the IMF. *Chicago Journal of International Law* 14, 377.

Polasik, M., Piotrowska, A., Wisniewski, T. P., Kotkowski, R., & Lightfoot, G. (2015). Price fluctuations and the use of Bitcoin: An empirical inquiry. *International Journal of Electronic Commerce* 20(1), 9–49.

Polasik, M., Piotrowska, A., Wisniewski, T., Kotkowski, R., & Lightfoot, G. (n.d.). Price fluctuations and the use of bitcoin: An empirical inquiry. *International Journal of Electronic Commerce* 20(1), 9–49.

Rani, S., Khang, A., Chauhan, M., & Kataria, A. (2021). IoT equipped intelligent distributed framework for smart healthcare systems, *Networking and Internet Architecture*. https://arxiv.org/abs/2110.04997v2, doi: 10.48550/arXiv.2110.04997

Sadeghi, R. (Ed.). (2013). *Financial Cryptography and Data Security*. Berlin Hiedelberg: Springer.

Sako, M. (1992). *Price, Quality and Trust: Inter-firm Relations in Britain and Japan*, Cambridge, Mass: Cambridge University Press.

Spyrou, Spyros (2013). Herding in financial markets: A review of the literature. *Review of Behavioral Finance* 5(2), 175–94.

Szabo (1998). Secure property titles with owner authority. Available at: http://nakamotoinstitute.org/secure-property-titles/ [10]

Szabo (2005). Bit Gold. Available at: http://unenumerated.blogspot.rs/2005/12/bit-gold.html

Tschorsch, & B. Scheuermann (2016). Bitcoin and beyond: A technical survey on decentralized digital currencies. *IEEE Communications Surveys & Tutorials* 18(3) (March), 2084–2123.

Weber, B. (2014). Bitcoin and the legitimacy crisis of money. *Cambridge Journal of Economics*. DOI:10.1093/cje/beu067

Yli-Huumo, J., Ko, D., Choi, S., Park, S., & K. Smolander (2016). Where Is current research on blockchain technology? A systematic review. *PloS ONE* 11(10), e0163477.

5 A Blockchain Approach to Improving Digital Linked Management Information Systems (MIS)

B. Akoramurthy, K. Dhivya, G. Vennira Selvi, and M. Prasad

CONTENTS

5.1	Introduction 62
5.2	Background 63
5.3	Information System and Blockchain Types..64
	5.3.1 Information Systems 64
	5.3.1.1 Information Creation..65
	5.3.1.2 Information Processing 65
	5.3.1.3 Information Analysis..65
	5.3.1.4 Information Visualization...65
	5.3.1.5 Digital Data Warehouse 65
	5.3.2 Public, Private, and Consortium Blockchains 66
	5.3.2.1 Public Blockchains...67
	5.3.2.2 Private Blockchains..67
5.4	Digital Block Information System (DIGBI) to DLIMS 2.0 68
5.5	Integration of MIS and Blockchain 70
5.6	Query Processing and Design of Smart Contract 71
	5.6.1 Smart Contract Algorithm Based on PINSI 72
	5.6.1.1 Initialization 72
	5.6.2 Algorithm PINSI 2.0 Cryptographic Part..72
	5.6.2.1 Encryption/Decryption Process 73
5.7	Blockchain-Based Transaction Design Process 73
5.8	System Implementation...76
5.9	System Evaluation 76
5.10	Performance Test...76
5.11	Performance Analysis..78
5.12	Conclusions 79
References...79

DOI: 10.1201/9781003269281-5

5.1 INTRODUCTION

Decision making is an art that cannot be learned instantaneously. Rather than the personal traits of the manager like risk-taking ability, time and change management, etc., lack of steadfast data is the major obstacle for any corporate manager in making decisions. Management of excess information has been a tricky challenge in identifying and utilizing accurate information at the correct time in various sectors.

Most companies face the maximum issues in dealing with information as their communication and flow among various departments leads to the major problem is stagnation and overhead of information leading to 'Information Silo' as description in article (B. Akoramurthy et al., 2020). In our previous work, a new framework was developed: DLIMS (Digital Linked Management Information System) for Management Information System (MIS), along with the PINSI (Purging INformation SIlo) algorithm, which purges the information silo with 80% efficiency.

But due to the emergence of many disruptive technologies such as blockchain, big data, and many others MIS has been integrated with newer technologies in order to address its limitations.

This chapter focuses on blockchain technology, which provides decentralization, trustk and immutability for the MIS. Since the data in a blockchain is not owned by a single entity, and the data is distributed among peers, the data is more secure and manageable and most important transparent.

The integration of blockchain technology to the proposed DLIMS system addresses the trust issue in the network as the immutable ledger is replicated and shared among members across the organization. Also, the work makes use of the biggest advantage of technology, such as creation and use of smart contracts that leads to real-time access.

This chapter emphasizes the integration of MIS with advanced metadata-based blockchain, which have the perfect characteristics and features for many upcoming applications. Though the integration framework sounds fantasized, some complications are found that have to be purged/tackled for efficient framework design.

Firstly, developing a suitable blockchain framework, and secondly with availability of various configurations and designs, it will be complicated to determine the best one. Lastly a complicated reconciliation process because of different languages in the IT environment leads to high error.

Thus, the research work includes:

(1) An improved blockchain information system based on DLIMS 2.0 is proposed.
(2) Also, the system has been upgraded with PINSI algorithm for not only purging information silo but also cryptographic functions that provide interfaces for blockchain integration.
(3) And, an information query system has been included as a new feature, which is powered by PINSI 2.0.
(4) Finally, a new smart contract was developed for the integrated MIS-Blockchain system.

5.2 BACKGROUND

The consortium blockchain and management information system has been analyzed for various sectors. In Mylrea & Gourisetti (2017), the system was introduced with Keyless Signature Infrastructure (KSI) and included a trust anchor between the chain nodes in the system. Bashir, Strickland, & Bohr (2016) talked about the mindset of people in using bitcoin as a currency and as a payment option in various applications.

Bhardwaj & Kaushik (2018) showed the blockchain in medical field applications, including sharing of medical records with doctors during treatment plans. Yin, Wen, Li, Zhang, & Jin (2018) emphasized a unique secured blockchain structure that is based upon lattice cryptography and provides a protective layer to safeguard from quantum attacks.

Sikorski, Haughton, & Kraft (2017) portrayed an information dissemination system based on blockchain for IoT applications (Khang et al., IoT & Healthcare, 2021).

Vandervort et al. (2015) describe the blockchain as a Service (BAAS) for public sector infrastructure events. Ølnes et al. (2016) analyzed the case of storing academic credentials in a blockchain-based wallet, which is also applicable for general case public.

Kshetri (2017) reported on security enhancement and credibility in supply chain management. Li and Wang (2018) researched the blockchain in a shared economy system with fair trading policy. In Chen (2018), analysis was done on a shared economy system security enhancement. Rahulamathavan et al. (2019) proposed a new payment mechanism that includes cryptocurrencies based upon privacy augmentation techniques.

Mödinger, Kopp, Kargl, & Hauck (2019) proposed a new protocol for broadcasting modifiable protection privacy system. Dang & Nguyen (2018) implemented a security mechanism for home automation. Tan, Zhou, Zhao, Zhao, Wang, Zhang et al. (2019) realized a sharing method of data among files on blockchain nodes.

Xia, Sifa, Asamoah et al. (2017) researched the data sharing systems in the medical industry and designed a schema for data sharing along with better access control mechanism using the blockchain.

Esposito, De Santis, Tortora, Chang, & Choo, 2018) studied the integrity of data in a cloud-based healthcare system. Liang, Shetty, Tosh et al. (2017) proposed an architecture for data availability and privacy for a cloud-based blockchain environment.

Chen, Zhang, Shi, Yan, & Ke (2018) did a comparison and performance analysis of a traditional database with blockchain, showing various performance testing parameters for blockchain technology.

Rather than application-oriented research, recently many authors have developed an open framework system blockchain technology. Yu, Li, Tian, & Liu (2018) proposed a framework for integration of edge computing with blockchain systems.

Yao & Wang (2018) analyzed the communication node in blockchain with the implementation authentication mechanism. Androulaki et al. (2018) reported on a hyperledger fabric permissioned blockchain that includes the tracking of unauthorized entry and request into the system and purging mechanism.

Le, Meng, Su, Yeo, & Thing (2018) implemented a forensic methodology for IoT which makes sure that all the trails in the system are recorded for a certain period of time to avoid the system trusting blindly.

Bartolucci, Bernat, & Joseph (2018) proposed a voting scheme, which is secret among the peer nodes in the blockchain, to identify legitimate users in the system, and also the mechanism makes sure that only authorized users can perform operations in the system with nominal fees.

These existing works show the possibilities of the application of blockchain technology to the upgrading of MIS. The advantages possessed by blockchain technology are that the digital information flowing through the system proposed will be able to be observed, allowing the information flow to be reorganized, which not only ensures the elimination information silo problem but also helps in better decision making.

5.3 INFORMATION SYSTEM AND BLOCKCHAIN TYPES

This section will discuss the differences between public, private, and consortium blockchain and various information systems.

5.3.1 INFORMATION SYSTEMS

Information is the cooked-up data and it is the mainstay for any kind of system. The drift or movement of any information in an organization is the bottom-line for communication. Information may be defined as follows.

Information is closed or structured if it has the following properties:

- unique identity
- integrity
- non-repudiation
- authenticity (when all the above properties are included) ✓ legitimacy

Old-style information systems were not computer based and were difficult to engage and manage. Finding particular information in that system was a tough job and no tracking of information was possible.

Moreover, it was a paper-based system, in which updating information needed more time and more resources, and there was even a possibility of having to restructure an entire work every time. For this reason, we moved on to computer-based information systems (Julie et al., 2020).

The composition of such a system will be able to provide all facilities, where all the operations – read, write, and update – can be done faster compared with traditional systems.

Due to the emergence of internet and mobile, the amount of data or information handled by computer-based systems is increasing every second. Every human produces data or information on a continual basis, even more than the corporate sector.

Because of this, current information systems require an extreme level of erudition to handle data dimensionality and automation for faster processing.

A Blockchain Approach to Improving Digital Linked MIS

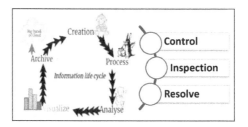

FIGURE 5.1 Information life cycle and its management.

Information systems evolved in a way such that they can work along with social networks. Information from social websites helps any organization in understanding market movements and customer needs.

Figure 5.1 illustrates the information life cycle.

5.3.1.1 Information Creation

Information creation is the easiest task or process in the lifecycle: since anyone can produce data through the use of social media and other avenues. Thus, corporations have information sources.

This evolution makes the information system more complex and increases the need for processing it.

5.3.1.2 Information Processing

As the size of the information acquired is of universal size, it needs to be processed at a faster rate of analysis or manipulation. Processing mostly involves removing unwanted data or information and finding data.

5.3.1.3 Information Analysis

Information analysis is a systematic process of interpreting information and discovering patterns in order to gain more insights in the information to improve information's business value. With the emergence of analysis software and its kind, handling analysis has become an easy task. Today's algorithms can handle information analysis in a simple manner.

5.3.1.4 Information Visualization

As we know already, a picture speaks more than words, and so the information analyzed must be visualized for better decision making. This is the most powerful medium of communicating information among employees in an organization.

5.3.1.5 Digital Data Warehouse

Information/data needs to be stored or archived in digital data warehouses or archives. However, recognizing information in a warehouse is a tedious task, which is why data mining concepts are needed.

Information also needs to be destroyed at times too. The complete lifecycle of information is the composition of an information system. Right from information creation to its usage, archival, or destruction it is the responsibility of the information system, which is now enhanced (DLIMS 2.0) to monitor the flow, utilization, and distribution of information along with security measures.

Modern information systems require information analysis and visualization for corporations to be able to understand customer demands, market trends, etc. Information destruction is yet another essential task since the rapidly evolving information once used needs to be destroyed in a secure manner. The need for a highly sophisticated and automated information system is increasing day by day.

5.3.2 Public, Private, and Consortium Blockchains

Information systems are categorized as centralized, decentralized, and distributed systems. In a centralized system, there is a solitary point of authorization and all the operations involved are controlled by a single end point. Most current internet applications are of this category. Also, client–server architecture is an emblematic centralized system architecture.

The decentralized system can be understood as no single point of contact system and no single point of failure. In other words, no individual system is in charge or controls any other systems. Each and every system in the network works independently and all the operations work in parallel mode.

Examples of typical decentralized system architectures are peer-to-peer and master–slave architectures. Disruptive technologies such as blockchain and cryptocurrency are also decentralized systems. Amazon and Google also follow the decentralized organization structure (Raikwar et al., 2019).

In the network, if the computation or control are shared among the group of servers, then the network is called a distributed one. Telephone and cellular networks and aircraft control systems are typical examples of distributed systems. Also, Facebook and Netflix (geographically distributed systems) are real-time examples of distributed systems.

Not every system comes under the above three categories. Some are centralized and distributed; other systems are decentralized and distributed. Many existing online services are run by centralized and distributed systems; while cryptocurrencies such as bitcoin and Ethereum, are decentralized and distributed systems.

Ethereum was designed as a public blockchain running in a fully decentralized network that is not controlled by any single entity and is exceptionally secured by using cryptographic and consensus algorithms such as proof-of-work and proof-of-stake. Private blockchains and consortium blockchains can be implemented on Hyperledger Fabric with configuration and/or access control. It is an open-source platform that is used to construct the distributed ledger system underlying a modular architecture that provides a high level of confidentiality, flexibility, resiliency, and scalability.

Many industries have started using the Hyperledger Fabric as a network architecture or a platform. It belongs to the category of private blockchain, which is powered by Linux.

5.3.2.1 Public Blockchains

Ethereum is a public blockchain in which any individual and organization can participate, and any computer can join the Ethereum network and become a node without permission required by any other node. Therefore, the public Ethereum is a permissionless blockchain where all nodes are considered equal, and no node is in control of any other node. These nodes form a peer-to-peer network. As a decentralized system, every node contains a copy of the blockchain that is synchronized with the entire transaction history.

As the number of nodes is large enough in the Ethereum network, the redundancy seems to be very high, and also it is possible to remove the node at any time without the flow of the operation in the Ethereum blockchain network. Blockchain enables trustless transactions without intermediaries such as clearinghouses, agents, or central organizations.

The blockchain network lets clients remain unidentified and perform transactions on a peer-to-peer basis. As a public ledger, the transaction data recorded on the Ethereum main chain is transparent to the public; and is immutable and cryptographically secured.

All records go through a mining process for verification before being added to the chain permanently. The computational cost involved in the mining process is very high, and the processing speed very slow. The mining process that runs a consensus algorithm to approve transactions is considered wasteful of energy and computing resources.

Currently, the Ethereum main chain can only process roughly 16 transactions per second. Due to this waste, the scalability is still the most challenging problem among the blockchain trilemma of trade-off among security, decentralization, and scalability.

There are three main types of blockchain transactions – send, approve, and read. The access privileges of these transactions are critical differentiators of public, private, and consortium blockchains. Although there are many different ways to build blockchain applications, these access privileges fundamentally define whether the blockchain is public, private, or consortium.

In a public blockchain, any node can send transactions to and read from the blockchain as shown in Table 5.1. Further, any node can participate in the consensus process to approve transactions.

Table 5.1 displays the traditional MIS with its parameters considered for its integration with blockchain technology.

5.3.2.2 Private Blockchains

The dogma of blockchain, being a permissionless and decentralized system that supports trust-based transactions and complete transparency, is fundamental to cryptocurrency and bitcoin. But this philosophy might get conceded in corporate sectors with the practical needs of reality.

Most government and business initiatives will not be attracted to the concept of transparency, i.e., they are not ready to share their business ledger records with other

TABLE 5.1
Traditional MIS with its Parameter Classification

Parameters	Management Information System (Social Web)
System Architecture	Centralized
Network Architecture	A New Five-Tier Client-server Model
Cloud computing service type	Network As a Platform (NAAP), Software as a Service (SAAS) & Infrastructure as a Service (IAAS)
Cloud type	Public
Front End	AngularJS, HTML, CSS, JavaScript, jQuery, SAAS, Flutter
Backend	Python, PHP, Java, SQL, C++, Node.js
Backend Frameworks	Django, Laravel, Spring, Express, Rails

dealings; but obviously want to keep their information confidential and private unless required for regulatory obligations or other specific purposes.

Many business managers might not want their competitors to partake in their blockchain transactions, but want to be the only entity with complete control of the blockchain, which is not possible since the system is decentralized. Therefore, private blockchains with modified features may be required to meet the needs of various enterprise applications.

These private blockchains can be integrated with management information systems for storage of secure and immutable vital records, to enable secure and trust-free transactions among internal departments, and to improve or lower the cost of a particular business process. The private blockchains can also play an essential role in being the single source of truth for key data.

The design and implementation of private blockchains (Khang et al., IoT & Healthcare, 2021) can take many forms, as a trade-off of a number of factors such as cost, speed, scalability, decentralization, transparency, choice of software, ease of implementation, and so on.

5.4 DIGITAL BLOCK INFORMATION SYSTEM (DIGBI) TO DLIMS 2.0

The Digital Block Information System (Digbi) is capable of handling modern information which is rapidly evolving with the help of blockchain technology. Data may be easily distributed among networked computers (nodes) that may be available in the same or different networks (Khang et al., AI & Blockchain, 2021).

As construction of blocks in a blockchain requires highly equipped computers, this information system will be capable of handling voluminous and rapidly evolving data. Data distributed over various blocks may be easily analyzed and visualized. Though the roots of this technology pertain to bitcoin, the influence of blockchain technology in the web world is still expanding (Khang et al., AI-Centric Smart City Ecosystem, 2022).

The proposed DIgbi aims to handle the issue of information silo and non-repudiation using blockchain technology. The utilization of blockchain technology helps in both identification of valid information for communication and ensures that non-repudiation is guaranteed.

The Digbi has been added with additional feature of managing its workflow by itself. The addition of management system emphasizes the control of the system process, information flow, block creation, consensus management, etc. This controlling enables the property of immutability, decentralization, and transparency throughout the system.

The concept of hashing is used in the system to recognize if tinkering is done to the information in the blockchain network. Almost all information in the block considered for hashing will be in the communication path; if any modification is detected, it can be identified easily since the information is in the chain code.

Moreover, the capability of holding voluminous information is also simplified with the use of hash wherein for an input of varying length, an output of a fixed length can be achieved.

DLIMS 2.0 provides the enactment of encryption standards and algorithms in the Hyper ledger Fabric network, including hashing, digital signature, verification, encryption and decryption, etc.

Figure 5.2 shows the information system module with security measures and management, where the PINSI plays a vital role in developing a smart contract and information query system.

The DLIMS2.0 module has provided the following benefit for the blockchain information system such as avalanche effect, if the data is modified in the block, it reflects back in the output of the system (see Figure 5.3).

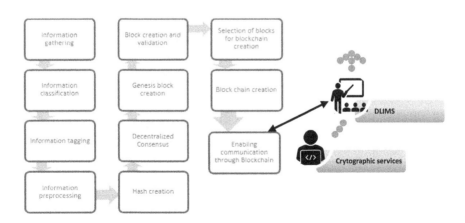

FIGURE 5.2   DLIMS 2.0 module.

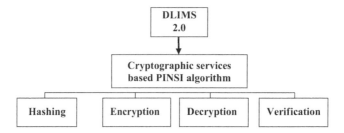

FIGURE 5.3 Digital Block Information System upgraded to DLIMS 2.0.

Source: DIGBI upgraded to DLIMS 2.0 diagram is designed by B. Akoramurthy, K. Dhivya, G. Vennira Selvi, and M. Prasad.

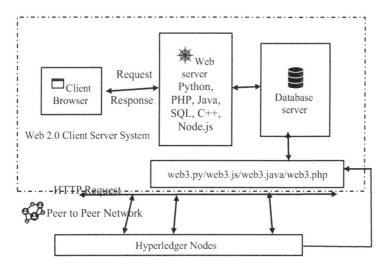

FIGURE 5.4 Integration framework.

Source: Integration Framework diagram is designed by B. Akoramurthy, K. Dhivya, G. Vennira Selvi, and M. Prasad.

5.5 INTEGRATION OF MIS AND BLOCKCHAIN

Integration of MIS and Blockchain is a hard-headed approach, when integrating current applications such as web 3.0 to conventional applications web 2.0 since web 3.0 is still maturing and moved to web 4.0. This kind of integration helps in solving many real-time problems or using applications (see Figure 5.4).

For instance, voters can be made to submit votes through a mobile platform (web 4.0) that is a blockchain-based system, in which the counting of votes can be done with certainty, since tampering is not possible in the system (Hanssen Rensaa et al., 2020).

Though there are choices of technologies and platforms available for construction of blockchain application, the proposed framework was developed based upon free open-source software in order to integrate the centralized architecture with decentralized architecture.

A Blockchain Approach to Improving Digital Linked MIS

The work follows a straightforward approach in integration process like data transfer between the two systems even though both are different architypes and almost all the operations involved in both the system are independent in nature which restricts the interactions between them. So, for every operation in the system an interface needs to be developed for smooth operation.

On the other hand, it is always simple to add a blockchain module to any business system.

5.6 QUERY PROCESSING AND DESIGN OF SMART CONTRACT

To query the user can login to the transaction information query system with the provided username and created password. Then the range of the query can be selected by the user after entering a unique ID or transaction code.

The result will be displayed as complete transaction information, which includes present transaction code, previous or last transaction, transaction time, amount of transaction, total time taken for the transaction, etc. This section covers the creation of a smart contract algorithm.

In order to improve query efficiency the smart contract was created based on PINSI. A greater number of users querying the system at the same time will give rise to congestion or deadlock in the system. A smart contract based on the policy will reduce the number of users accessing the system at the same time.

Every user in the network will get a range of access rights from the smart contract algorithm. Only core users in the network will be able to know the historical information about the transaction and everything else, whereas non-core users know only the current status of the transaction.

The system checks whether the user has logged-in as admin, and depends on the context; the system will invoke the smart contract to query the transaction information through a unique ID or transaction code as seen in Figure 5.5.

```
Check if Login User is Admin, then
    Input: string UniqueID/Aadhar Number
    GetState function () //returns the current state of the transaction
    Unmarshal (data [] byte, v interface {}) error // for decoding the string input
    PINSI hash operation on x ⟶ X
    CouchDB -PINSI hash operation on Y
    Hash Result comparison X & Y
    If X==Y no data is modified
    Display historical item ()
else if Login user is not admin, then
    Input: string UniqueID/Aadhar Number
    Display only current status and transaction code
    end if
    return UniqueID
```

FIGURE 5.5 Smart contract algorithm.

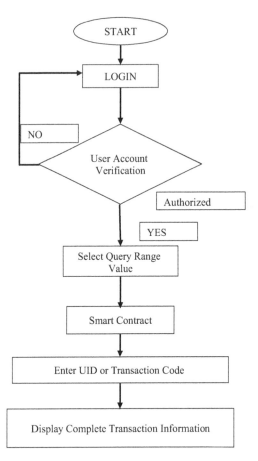

FIGURE 5.6 Flowchart of query process in the transaction.

5.6.1 Smart Contract Algorithm Based on PINSI

5.6.1.1 Initialization

Figure 5.6 shows the flowchart for the query process.

5.6.2 Algorithm PINSI 2.0 Cryptographic Part

Initialization ()
Private key generation of User x:
$D_x \in [1, n-2]$ // Random number generated by the User X.
Public key computation of User x:
$P_x = D_x \cdot B$ // B is the base point of the curve
Public/Private key pair of user x:
$PPP = (P_x, D_x)$

5.6.2.1 Encryption/Decryption Process

Let us assume that the plain text or a string is encrypted as "E" and its length is represented as "len."

User x has to do the following:

```
Encryption ()

k ∈ [1, n-2]//Random number generation
Ec₁= [k]. B (a₁, b₁)//ecliptic's curve point calculation
M= [h]. Pₓ //point calculation in elliptic curve
Ec₂ =[k]. Pₓ = (a₂, b₂)// ecliptic's curve point calculation
S= KD (a₂||b₂, len)// calculation of Key derivation function
Ec₂= S ⊕E;
Ec₃ = hash(a₂||E||b₂);
C= Ec₁||Ec₂||Ec₃;// cipher text output
```

```
Decryption ()

Pick the string bit Ec₁ from C;
M= [h]. Ec₁ // ecliptic's curve calculation
[Dₓ]. Ec₁ = (a₂, b₂);
S= KD (a₂||b₂, len);
Pick the string bit Ec₂ from C;
Calculate E' = Ec₂ ⊕ S;
w= hash(a₂||E'||b₂); Pick the string bit Ec₃ from C;
E' is the plain text after the decryption process.
```

5.7 BLOCKCHAIN-BASED TRANSACTION DESIGN PROCESS

AI-based systems are classified into four types of machines: reactive AI machines, limited memory AI machines, theory of mind AI machines, and self-aware AI machines.

The following section shows the transaction design process based on the blockchain as shown in Figure 5.7.

From the definition of consortium, the network along with its users, channels, peers, and ordering nodes and services are created. The configuration of the channels includes the identity of the client, membership ID, and chain code ID, and channel policies find the ordering service, which can be called administrator inside the network. In the creation genesis block, which is called the first block in the blockchain, the prime step is the creation of the consortium. A consortium, which is defined by the organization itself, is a composition of two or more organizations in the network of the blockchain (Aitzhan & Svetinovic, 2018).

Based on the channel policy of the network, any transactions can take place within and among the organizations. In order to connect various components present in the client-side applications or even in the network, channels are required.

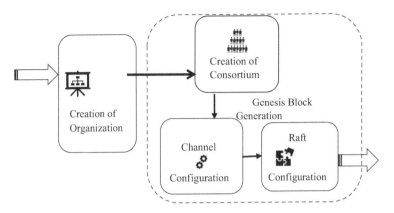

FIGURE 5.7 Organization creation for transaction design process.

The creation of channels is based on the ordering service configuration block in the network. This configuration block is also responsible for assessing the validity of the created channel. Creation of a channel in the network also has the advantage of isolating data and shielding it in a secured manner. Any transactions in the organization should be authorized by the channel in order to have interaction with the process (Efthymios et al., 2022).

Governance of a channel is defined by the policies within itself. Altogether, channel creation and its configuration is called genesis block generation since it is the first block in the blockchain network. It is also sometimes called antecedent block in the network.

The internal block structure of the genesis block will be known to all the other nodes in the chain. After carefully collecting information required for communication in the network, the genesis block will be created. Once the information in the block is encoded using cryptographic algorithms, no modification can be done, which makes the data immutable.

If any of the information in the blockchain is to be accessed, then all the other nodes will be notified for security reasons and information cannot be tampered with. In case any modification has to be done in the block, approval has to be given by consensus algorithm as shown in Figure 5.8.

Figure 5.8 shows the organizational flow that takes place in the transactional design process. Here the first step is to create a certificate authority that provides certificate services to the blockchain users. Services like new user admission, invoking transactions in the blockchain, and establishing secured connections between the users of the network.

Figure 5.9 exemplifies a transaction process based on DLIMS 2.0. Since the transaction is going to take place in the permissioned blockchain network, the member needs to have an identity issued by the Certified Authority, which supplements trust within the network to participate in the transaction process.

To make the proposed DLIMS system more manageable, as a primary step in the transaction process, the Managed Service Provider (MSP) is created and integrated

A Blockchain Approach to Improving Digital Linked MIS

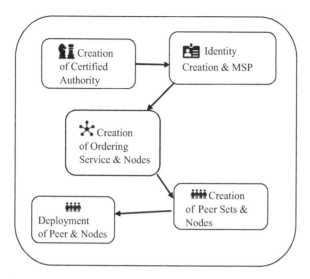

FIGURE 5.8 Organizational flow for transaction design process.

to manage and enact smart contracts. In the defined organization, members are issued their own private key and certificates, which are created and managed by the MSP.

Members will be verified based on their identity and their roles in the organization.

In the next step the buyer/user sends a PROPOSE Message to invoke a transaction. The NODE.js SDK prepares a PROPOSE message in the format as follows:

$$<PROPOSE, t_x, [anchor]>$$

anchor is optional

t_x is the set of identities, payload and timestamp. i.e., $t_x = \{UID + CCID+TPL+TS+ USIG\}$

where

- UID is a digital block ID of the user
- CCID is the code of block in network to which the transaction relates
- TPL is the transaction payload
- TS is a timestamp or a unique serial number stored in each block
- USIG is a user signature on the transaction fields

Upon receiving the transaction request, the endorsing peer will validate the message in the following way:

- Checks the message is well-versed or not
- Checks the message is submitted the first time
- Signature verification
- Checks the user authorized or not in the organization to carry out transaction

Every endorsing peer invokes the chain code function by passing the transaction proposal as an argument. This produces a transaction result with response value and RW sets.

5.8 SYSTEM IMPLEMENTATION

In order to provide a more secured system, the DLIMS has been combined with Hyperledger Fabric, making it super DLIMS 2.0. The process involved in design and implementation of the system includes upgrading the PINSI algorithm along with the permission management algorithm in the DLIMS 2.0 module and the generation of a smart contract based on the network as shown in Figure 5.9.

The following is the architecture of the information query system based on DLIMS 2.0.

5.9 SYSTEM EVALUATION

The developed system was executed on the test bench configured with an Intel Core i5 processor, 8 GB of memory, and Windows 11 64-bit operating system. The system also runs on a MintOSV1.4.0 (64-bit) virtual machine.

5.10 PERFORMANCE TEST

The system handled transaction records of range 10T-3000T during the system test. Table 5.2 exhibits the overall time taken (elapsed time) and average time for

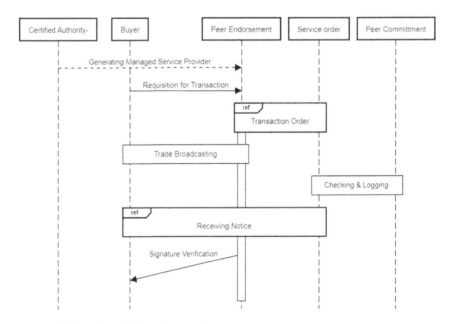

FIGURE 5.9 DLIMS 2.0-based transaction process.

every transaction query information. When 4000 transaction records are queried, the average time in the system is only 30.210ms per query provided all the queries are processed successfully as shown in Figure 5.10.

Table 5.2 shows the blockchain information query system achieved.

The graph in Figure 5.11 shows the relationship between elapsed time and the average time taken for each query in the transaction process. From the graph, we can conclude that the average query time is indirectly proportional to the number of query information transactions.

$$i.e., \text{Average Query time } \alpha \frac{1}{\text{Amount of Query Information in Transaction}}$$

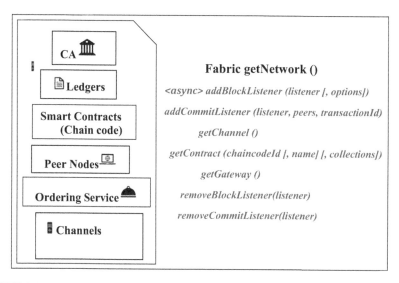

FIGURE 5.10 Hyperledger Fabric information query system.

Source: The diagram of Hyperledger Fabric Information Query System is designed by B. Akoramurthy, K. Dhivya, G. Vennira Selvi, M. Prasad.

TABLE 5.2
Query Performance Test

Amount of Query	10	100	200	500	1000	2000	3000	4000
Elapsed Time (ms)	794.1	3770.8	8756.3	19345.57	36542.1	52987.2	78925.5	9923.5
Average time per Query (ms)	56.16	52.11	52.10	50.65	48.14	41.56	33.93	30.21

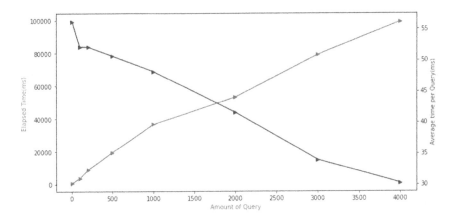

FIGURE 5.11 Elapsed time vs. average time per query vs. amount of query.

Source: The chart was designed by B. Akoramurthy, K. Dhivya, G. Vennira Selvi, and M. Prasad.

The good performance is that due to the implementation of new intelligent smart contract in the system, where the algorithm hides the data and related things from the members of the organization, which makes members only query the network rather than accessing directly. Also, the size of the members in the network is more than that of the admins of the network. As the core data becomes stealthy from members, the load calculation of the information system is reduced along with the amount of query data (see Figure 5.11).

5.11 PERFORMANCE ANALYSIS

In terms of security, the developed system from the PINSI algorithm provides an extra layer of secureness around the network, by means of digital signatures. And the algorithm ensures the authenticity and legitimacy of the message from the client.

Because of this, the system becomes the most trusted one among the organizations. Also, the algorithm lets the user or sender generate a digital signature on the data produced and lets peers verify the reliability of the data sent by the sender in the network (Ma et al., 2019).

If a malevolent attacker tries to steal the identity of a particular node in the network, he/she must kiln the certificate and identity from the certificate authority, which is impracticable because of the algorithm. This means that the identity created in the system is more trustworthy.

Currently, some research results are available on the enhancement of smart contract and crypto algorithms used in the consortium blockchain. In order to show the best, we are comparing our system performance with the following research achievements in the same field.

Liang (2017) displays public blockchain information system integrated with cloud computing. Chen (2018) reveals the private blockchain information system

TABLE 5.3
System Performance Comparison

Research Work	Type of Blockchain	Conditional Query	Algorithms	Average time per Query(ms)
X. Liang et al., 2017)	Public blockchain	NO	Blockchain Receipt Validation Algorithm	221
(S. Chen et al., 2018)	Consortium Blockchain	NO	SM2 & SM3	125
T. Yan et al., 2019)	Hyperledger Fabric	YES	The original	49.51
(B. Ahmed et al., 2019)	Hyperledger Fabric	NO	Consensus	40.23
DLIMS 2.0	**Hyperledger Fabric**	**YES**	**PINSI**	**30.210**

performance. Yan (2019) shows the conditional query performance on the system and Ahmed (2019) tested the performance of permissioned blockchain network (Bhambri et al., Cloud & IoT, 2022).

Results shows that DLIMS 2.0 has better performance on the network and has high query efficiency, due to the PINSI algorithm and smart contract.

Table 5.3 shows the comparison with other research works defined with DLIMS 2.0 having better average query time.

5.12 CONCLUSIONS

The integration of blockchain technology with traditional MIS shows the single source information truth across the enterprises or even in the supply chain management system (Abeyratne & Monfared, 2016).

The MIS is upgraded with immutability where no modification or tampering of information can be done. Since the blockchain has some security concerns, a new smart contract algorithm based on PINSI helped to resolve the issue.

In this chapter we implemented the private and consortium blockchains on the platform Hyperledger Fabric. As a new feature to DLIMS, a transaction information query was developed as a prototype, in which after being examined, the average time for a query in the transaction process is only 30.210 ms in the blockchain network.

REFERENCES

Abeyratne, S. & Monfared, R., Blockchain ready manufacturing supply chain using distributed ledger. *Int. J. Res. Eng. Technol.* 2016, 5, 1–10. doi: 10.15623/ijret.2016.0509001

Ahmed, B., R. Laura, H.M. Qusay, E. Khalid, & C.K.H.A. Patrick, A permissioned blockchain-based system for verification of academic records, in *Proc. 2019 10th IFIP International Conference on New Technologies, Mobility and Security (NTMS)*, 2019, pp. 1–5.

Aitzhan, N. Z., & D. Svetinovic, Security and privacy in decentralized energy trading through multi-signatures, blockchain and anonymous messaging streams, *IEEE Transactions on Dependable and Secure Computing*, 15(5), 840852, Oct. 2018.

Akoramurthy, B., & T. Ananth Kumar, *Digital Linked Information System Using Blockchain Technology-Overwhelming Information Silo, Blockchain Technology: Fundamentals, Applications, and Case Studies* (1st ed.). CRC Press, 2020. https://doi.org/10.1201/9781003004998

Androulaki, E. et al., Hyperledger fabric: A distributed operating system for permissioned blockchains. In *Proceedings of EuroSys 2018 Thirteenth EuroSys Conference*, 2018, pp. 15–30.

Bartolucci, S., P. Bernat, & D. Joseph, SHARVOT: secret SHARe-based VOTing on the blockchain. In *Proceedings of ACM/IEEE 1st International Workshop on Emerging Trends in Software Engineering for Blockchain*, 2018, pp. 30–4.

Bashir, M., B. Strickland, & J. Bohr, What motivates people to use bitcoin? In *Proc. SocInfo 2016*, 2016, vol. 10047, Springer, pp. 347–67.

Bhambri, Pankaj, Sita Rani, Gaurav Gupta, & Khang, A., *Cloud and Fog Computing Platforms for Internet of Things*. Chapman & Hall. ISBN: 9781032101507, 2022.

Bhardwaj, S., & M. Kaushik, Blockchain – technology to drive the future. In *Proc. Smart Computing and Informatics*. 2018, 78, Springer, pp. 263–71.

Chen, S., J. Zhang, R. Shi, J. Yan, & Q. Ke, A comparative testing on performance of blockchain and relational database: Foundation for applying smart technology into current business systems. In *Proc. Distributed, Ambient and Pervasive Interactions: Understanding Humans*, 2018, 10921, pp. 21–34.

Dang, T.L.N., and M.S. Nguyen, An approach to data privacy in smart home using blockchain technology. In *Proc. 2018 International Conference on Advanced Computing and Applications (ACOMP)*, 2018, pp. 58–64.

Efthymios et al., Using blockchain and semantic web technologies for the implementation of smart contracts between individuals and health insurance organizations, *Blockchain: Research and Applications* 3(2), 2022, 100049, ISSN 2096-7209, https://doi.org/10.1016/j.bcra.2021.100049.

Esposito, C., A. De Santis, G. Tortora, H. Chang, & K.-K.R.Choo, Blockchain: a panacea for healthcare cloud-based data security and privacy? *IEEE Cloud Computing* 5(1), 31–7, Mar. 2018.

Hanssen Rensaa, Jens-Andreas, Danilo Gligoroski, Katina Kralevska, et al., VerifyMed – A blockchain platform for transparent trust in virtualized healthcare: Proof-of-concept. In *Proc. of 2020 2nd International Electronics Communication Conference*, 2020, pp. 73–80.

Julie, E.G., Vedha Nayahi, J.J., & Jhanjhi, N.Z. (Eds.). (2020). *Blockchain Technology: Fundamentals, Applications, and Case Studies* (1st ed.). CRC Press. https://doi.org/10.1201/9781003004998

Khang, A., Geeta Rana, Ravindra Sharma, Alok Kumar Goel, & Ashok Kumar Dubey, The role of artificial intelligence in blockchain applications, *Reinventing Manufacturing and Business Processes Through Artificial Intelligence*, 2021, doi: 10.1201/9781003145011.

Khang, A., Rani, S., & Sivaraman, A.K. (Eds.). (2022). *AI-Centric Smart City Ecosystems: Technologies, Design and Implementation* (1st ed.). CRC Press. https://doi.org/10.1201/9781003252542.

Kshetri, N. Blockchain's roles in strengthening cybersecurity and protecting privacy, *Telecommun. Policy* 41, 1027–38, Dec. 2017.

Le, D.P.H., Meng, L. Su, S. L. Yeo, & V. Thing, BIFF: a blockchain-based IoT forensics framework with identity privacy, in *Proceedings of TENCON 2018*, 2018, pp. 2372–7.

Li, B., & Y. Wang, RZKPB: a privacy-preserving blockchain-based fair transaction method for sharing economy. In *Proc. 2018 17th IEEE International Conference on Trust, Security and Privacy in Computing and Communications/12th IEEE International Conference on Big Data Science and Engineering*, 2018, pp. 1164–9.

Li, B., Y. Wang, et al., FPPB: a fast and privacy preserving method based on the permissioned blockchain for fair transactions in sharing economy. In *Proc. 2018 17th IEEE International Conference on Trust, Security and Privacy in Computing and Communications/12th IEEE International Conference on Big Data Science and Engineering*, 2018, pp. 1368–73.

Liang, X., S. Shetty, D. Tosh, C. Kamhoua, K. Kwiat, & L. Njilla, ProvChain: A blockchain-based data provenance architecture in cloud environment with enhanced privacy and availability. In *Proc. 2017 17th IEEE/ACM International Symposium on Cluster, Cloud and Grid Computing (CCGRID)*. IEEE, 2017, pp. 468–77.

Ma, C., X. Kong, Q. Lan, et al., The privacy protection mechanism of hyperledger fabric and its application in supply chain finance, *Cybersecur* 2(5), Jan. 2019.

Miller, D., Blockchain and the Internet of Things in the industrial sector, *IT Professional* 20(3), 5–18, May/June 2018.

Mödinger, D., H. Kopp, F. Kargl, & F.J. Hauck, A flexible network approach to privacy of blockchain transactions. In *Proc. 2018 IEEE 38th International Conference on Distributed Computing Systems*, 2018, pp. 1486–91.

Mylrea, M., & S.N.G. Gourisetti, Blockchain for smart grid resilience: exchanging distributed energy at speed, scale and security. In *Proceedings of 2017 Resilience Week*, 2017, pp. 18–23.

Nathan, S., P. Thakkar, & B. Vishwanathan, Performance benchmarking and optimizing hyperledger fabric blockchain platform, in *Proc. 2018 IEEE 26th International Symposium on Modeling, Analysis, and Simulation of Computer and Telecommunication Systems (MASCOTS)*, 2018, pp. 264–76.

Ølnes, S., Beyond bitcoin enabling smart government using blockchain technology. In *Proc. EGOVIS 2016*, 2016, vol. 9820, Springer, pp. 253–64.

Rahulamathavan, Y., R.C.-W. Phan, M. Rajarajan, S. Misra, & A. Kondoz, Toward privacy and regulation in blockchain-based Cryptocurrencies, *IEEE Network* 33(5), 111–17, Oct. 2019.

Raikwar, M., D. Gligoroski, & K. Kralevska. SoK of used cryptography in blockchain. *IEEE Access*, 2019, 148550–75.

Rani, Sita, Khang, A., Meetali Chauhan, & Aman Kataria, IoT equipped intelligent distributed framework for smart healthcare systems, *Networking and Internet Architecture*, 2021, https://arxiv.org/abs/2110.04997v2, doi: 10.48550/arXiv.2110.04997

Sikorski, J., J. Haughton, & M. Kraft, Blockchain technology in the chemical industry: machine-to-machine electricity market, *Appl. Energy* 195, 234–46, June 2017.

Tan, H.. T. Zhou, H. Zhao, Z. Zhao, W. Wang, & Z. Zhang, et al., File data protection and sharing method based on blockchain, *J. Softw*. 30(9), 1–15, Apr. 2019.

Vandervort, D., D. Gaucas, & R. S. Jacques, Issues in designing a bitcoin-like community currency. In *Proc. FC 2015*, 2015, vol. 8976, Springer, pp. 78–91.

Xia, Q., E.B. Sifa, & K.O. Asamoah, et al., MeDShare: trust-less medical data sharing among cloud service providers via blockchain, *IEEE Access* 5(99), 14757–67, July 2017.

Yan, T., W. Chen, P. Zhao, Z. Li, A. Liu, & L. Zhao, Handling conditional queries on hyperledger fabric efficiently. In *Proc. International Conference on Web Information Systems Engineering*, 2019, pp. 48–62.

Yao, H., & C. Wang, A novel blockchain-based authenticated key exchange protocol and its applications. In *Proceedings of IEEE 3rd International Conference on Data Science in Cyberspace*, 2018, pp. 609–14.

6 A Perspective on Blockchain-based Cryptocurrency to Boost A Futuristic Digital Economy

S. Shivam Gupta, Sarishma Dangi, and Sachin Sharma

CONTENTS

6.1	Introduction	83
6.2	Related Work	86
6.3	Background	86
	6.3.1 Smart Contract	88
	6.3.2 Decentralized Finance	89
	6.3.3 Digital Economy	89
6.4	Current Market Scenario	90
6.5	Case Study: India	93
6.6	Application Areas	95
	6.6.1 Financial Services	95
	6.6.2 Government Sector	95
	6.6.3 Health Sector	95
	6.6.4 Entertainment Sector	96
	6.6.5 Digital Identification	96
6.7	Challenges	97
6.8	Conclusion	98
References		98

6.1 INTRODUCTION

For ages, the barter system was used as a medium of exchange for any kind of service or product. Throughout the 18th century retailers begin to abandon the usual system of bartering, as there was no common measure of value (Humphrey, 1985). This system evolved to become more efficient and effective, with the use of paper money.

The country's central bank or treasury took charge of printing and circulating money. Paper money was useful, safe, reliable, nonvolatile, and most importantly it provided a common measure of value. It was backed up by the government and soon, it became the backbone of the whole financial sector.

Over decades, everything seemed to be in place until "The Great Recession" in 2008 that blew the financial systems of many countries of the world. This incident unveils the shortcoming of the banking system such as interest rate manipulation, insufficient regulation, toxic mortgages but corruption has always been a significant one (Business Insider India, 2020). This tragedy shook people's faith in banking so much that a new class of assets came into being, which doesn't have the backing of any formal bank (Combating Corruption, 2022).

Cryptocurrency eliminates the central authority which was the prominent reason for the crisis. The cryptocurrency used the concept of decentralization, which can be used as an alternative to traditional currency or paper currency. Bitcoin was the first cryptocurrency introduced in November 2008 based on a white paper by Satoshi Nakamoto (Bitcoin_Whitepaper_Document_HD), which laid the basic foundation of other cryptocurrencies founded later on.

The primary technology on which cryptocurrencies are based is termed as "*blockchain*." It is a distributed digital ledger that is immutable, and stores data in a decentralized manner.

Blockchain technology has been rising these past years as data privacy is of everyone's concern now, and this can provide us the same with no invasion of the third party in between. In 2021, the worldwide expenditure on blockchain based solutions was predicted to reach $6.6 billion.

Further, studies have shown that expenditure on blockchain based solutions for real-life issues will continue its growth in the coming years, with an estimate to reach $19 billion by 2024 (Global Spending on Blockchain Solutions, 2021).

As soon as blockchain came to light, developers and economists began to harness this technology for many different reasons, such as providing the world with a better and safer platform to make transactions. In less than a decade, open-source organizations like DAO (Decentralized Autonomous Organization) were launched in the market to deliver a novel decentralized business model used for establishing commercial enterprises (Swinburne News, 2016).

Many private organizations are also working on a large scale toward blockchain development like Hyperledger, established in 2015 by the Linux Foundation (About – Hyperledger Foundation, 2022). Hyperledger is an umbrella project developed as open-source blockchains and related tools. It was established to provision the collective development of blockchain based distributed ledgers. Azure blockchain workbench provides various Azure services and helps the user to design and create blockchain-based applications (Microsoft Docs, 2022).

Many countries such as Sweden, El Salvador, and Switzerland accepted blockchain technology in one way or the other and adopted the same in their governing sector. For instance, Switzerland accepted bitcoin as its legal tender and installed 100+ bitcoin ATMs to encourage its citizens to use cryptocurrency and build trust toward it. Independence from any central authority, security and automation are some of the reasons countries are eager to embrace this technology.

FIGURE 6.1 Application blockchain.

Source: Diagram is designed by S. Shivam Gupta, Sarishma Dangi, and Sachin Sharma.

Although the benefits are huge in adopting cryptocurrency, there are some limitations such as scalability, governance, privacy, and energy consumption that can't be ignored. Due to these factors, many countries are hesitating to adopt blockchain including India.

While blockchain developers are working toward reducing these limitations, many businesses have emerged, solving real-life problems such as MedicalChain Homeland Security, 2021). It is a blockchain-based platform that stores a record of patients in a ledger, which can be accessed by doctors or hospitals for reference in ongoing treatment.

Blockchain is a versatile technology that can be used in various sectors such as retail (Chakrabarti et al., 2017), eCommerce (Kumar et al., 2020), banking (Hassani et al., 2018), digital art (Zeilinger et al., 2016), healthcare (Tomar et al., 2020), and supply chain management (Queiroz et al., 2020) (see Figure 6.1).

The key contributions unique to this work are outlined as follows:

- To identify the current market scenario globally of the Indian digital economy
- A case study on India is presented
- To discuss the various application areas for the adoption of cryptocurrency in the digital economy along with real-life business use cases
- To identify the challenges while adoption as well as the implementation of blockchain-based cryptocurrency in different sectors.

The remaining chapter is structured as follows:

- Section 6.2 provides the related work in the same area.
- Section 6.3 provides the technical background of underlying technology which are boosting the digital economy.
- Section 6.4 discusses a current market scenario in regards to blockchain and cryptocurrency.
- Section 6.5 gives an overview of the market scenario of India in terms of blockchain involvement in the economic sector.
- Section 6.6 is focused on different sectors where blockchain can be implemented along with some real-life business cases.
- Section 6/7 highlights the major challenges countries have faced while integrating blockchain into their system followed by the conclusion of this work.

6.2 RELATED WORK

In this section, we discuss the related work being done in the area of cryptocurrency adoption to boost the digital economy of various countries across the world. In recent studies on the blockchain, B. A. V. et al. (2017) worked on how to cope with different security threats, proposed a low-cost and safe smart grid system authentication method using a blockchain. According to Christan's research (Catalini, 2022), they introduced blockchain technology and reorganizes the current centralized regulation system.

According to Marc Pilkington's study (Pilkington et al., 2016), work presented the core concepts of blockchain and discussed the potential risks and drawbacks of public distributed ledgers and the shift toward hybrid solutions. In the work (Deshmukh et al., 2021), the state of the Indian digital economy was analyzed however, the work is heavily focused on e-commerce and fails to address other sectors of the economy, which is the key focus of our work.

In the work (Maxmudjanovna et al., 2020), the underlying potential of blockchain to develop the digital economy was discussed. The work is limited to theoretical prospects only. Our work provides real-life business case studies with an overview of the current market scenario on a global scale, with a separate discussion on the market scenario of the Indian digital economy.

6.3 BACKGROUND

In this section, we discuss the underlying concepts related to cryptocurrency. The traditional centralized system can be defined as a major hub or hubs with which peers communicate. This centralized system came with major flaws, such as no possibility of data recovery, if the server node fails and there is no backup, we'll lose all the data right away.

Another major flaw in a centralized system is the dependence on network connectivity, i.e. if the node losses network connectivity; the whole system goes down as there is only one central node. Energy consumption and data privacy also have share of the list of shortcomings. These flaws restricted us in many ways, which eventually led to the need for a decentralized and more transparent system. Thus, blockchain emerged as an answer to them.

Decentralization refers to the simple transfer of control as well as decision-making away from a centralized authority and toward a more distributed and a decentralized network. The decentralized networks thrive to minimize the level of security and trust that users have to place in each other. Decentralization has noteworthy benefits over-centralization such as it provides a trust-less environment, so no one has to know or trust anyone else; it improves data reconciliation, to provide every entity real-time data; it deduces points of weakness as every node in the network have exact copy of the ledger (What is Decentralization?, 2020).

Blockchain provides a distributed and transparent ledger that ensures the integrity of any document. Blockchain is a decentralized digital ledger that stores the origin of a digital asset. The data in the ledger is distributed which makes the record immutable. This data is decentralized, allowing instantaneous access to the public. By characteristic design, the stored data on a blockchain is immutable, which makes it a genuine disruptor for industries like payments, healthcare, and banking (What Is Blockchain Technology?, 2022). The most prominent and noteworthy application of blockchain is cryptocurrency (Khang et al., AI & Blockchain, 2021).

Cryptocurrency: The term cryptocurrency is a combination of the word "crypto" meaning cryptography, which is used to ensure all transactions sent between users are encrypted. The word "currency" means a medium of exchange. Cryptocurrency is a digital payment alternative that eliminates the role of banks to verify every transaction (Cryptocurrency Bill, 2022). Its system allows us to send payments across the globe with the help of its peer-to-peer network.

Cryptocurrency changes the whole dynamic of transferring money, as physical presence is not required, unlike traditional currency as Table 6.1. A secure ledger records all the essential details of payment and is open for the public to verify (Free Crypto Screener, 2022). Some of the cryptocurrencies prominently used in the industry are mentioned in Table 6.1.

Table 6.1 displays some of the cryptocurrencies available in the market.

Many onsets of crypto exchanges have emerged for retail investors. This sector gained more popularity since the 2019 global pandemic, one such Indian exchange is Coinswitch (Beck, 2019). Coinswitch changed the whole experience for people in India toward investing in crypto with its easy user interface. Coinswitch recently became a unicorn, after raising $260 million from its new investors (Forbes India, 2022), raising the bars for its competitors in the market (see Figure 6.2).

TABLE 6.1
Cryptocurrencies

Cryptocurrency	Creator	Market Capital	Year of launch	Symbol
Bitcoin	Satoshi Nakamoto	₹58.83T	2008	BTC
Ethereum	Vitalik Buterin	₹28.34T	2013	ETH
Litecoin	Charlie Lee	₹679.26B	2011	LTC
Uniswap	Hayden Adams	₹751.61B	2018	UNI
Dai	MakerDAO	₹692.22B	2017	DAI
Cardano	Charles Hoskinson	₹2.68T	2017	ADA
Terra	Terraform Labs	₹1.99T	2018	LUNA
Tron	Justin Sun	₹419.83B	2018	TRX
Basic Attention Coin	Brendan Eich	₹80.97B	2017	BAT
Decentraland	Esteban Ordano	₹274.19B	2020	MANA

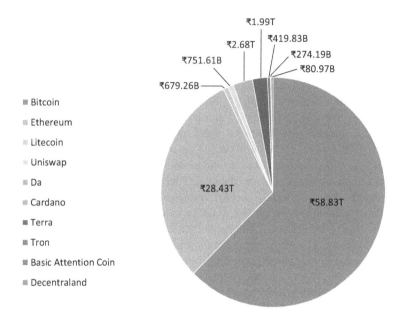

FIGURE 6.2 Capitals of cryptocurrencies.

Source: Chart is designed by S. Shivam Gupta, Sarishma Dangi, and Sachin Sharma.

Apart from cryptocurrency, another application that holds the major significance and differentiates blockchain from traditional technology is *Smart Contracts*.

6.3.1 SMART CONTRACT

A smart contract is a self-executable program code that automatically enforceable itself on the top of blockchain network to manage difficult business logic. It states the terms of an agreement, designed for the user, lies among untrustworthy parties. The logic used in smart contracts is cryptography, that enables the blockchain network to provide trust and authority to all required parties in transactions.

Smart contracts designed on the simple logic of "if/when…then…" A chain of computers executes the actions when stated conditions have been fulfilled and verified successfully. Some of these actions could include managing finances such as release of funds to the concerned parties, issuing a ticket, sending notifications, or registering a vehicle.

The blockchain is automatically updated as and when the transactions are completed. The stored transactions are immutable, and only the concerned parties that have been granted permission can see the results, which makes smart contracts trustworthy (Swedish Blockchain Association, 2022).

Smart contracts are versatile due to which they can be beneficial in smoothening the workflow and improving the overall operations for any organization. Smart contracts provide us accuracy, they execute themselves without any scope of error when all the conditions are fulfilled.

Transparency is also one of the features smart contracts provide, all the transactions are completely public and are accessible to all relevant parties. This feature leaves no space for miscommunications and can immensely bring down the loss due to gaps in communication.

With such clear communication and accuracy, speed in operations is complimentary. Smart contracts can save hours of manual labor with automation. As in smart contracts, all the data is encrypted which makes it one of the most secure items across the web. Organizations can go paper-free with the embedment of smart contracts in their operations.

Being paper-free will not only improve efficiency but also impact the environment in a positive manner. In small or medium scale businesses, data retrieval is one of the most painful procedures due to a lack of resources. Smart contracts also provide this facility of data backup, as it stores all the data online, and can retrieve data at any given time.

6.3.2 Decentralized Finance

"Finance" is a set of blockchain-based financial service that intend to replace the traditional finance system. As of now the system is comparatively referred to as "Centralized Finance."

When DeFi is compared to the traditional services, it comes with numerous benefits as its ecosystems matures, deploying a financial application will get much less complex and demanding. A real-life use case is the Ethereum blockchain, which reduces the operational costs and lowers entry barriers (Statista, 2021).

DeFi does not depend on financial intermediations such as brokerage, exchanges, or banks to offer traditional financial services, and instead uses smart contracts on blockchains network. It targets anonymity, security, and ease head-on. The three major functions of DeFi outlined as follows:

- Translating monetary banking services i.e. issuance of stablecoins. The stablecoins are a type of cryptocurrency that is intended to maintain a stable market. The idea behind stablecoins is to incorporate some of the plus points of both fiat currency and cryptocurrency.
- Providing peer-to-peer lending & borrowing platforms is one of the most fundamental category of DeFi. Anyone can deposit their crypto in the smart contracts as collateral and can borrow against it. Its system automatically matches the lenders and borrowers, and adjust the interest rate concurrently, based on basic demand and supply techniques.
- Enabling advanced financial services such as decentralized exchanges, tokenization platforms, predictions markets, and derivatives.

6.3.3 Digital Economy

The global economy is the accumulation of activities that are conducted both within a country, and between the different countries. This globalization of the economy enhances the technology and thus the development of each country due to the presence of a competitive market (Digital Economy Compass, 2020). The term digital economy was coined in the year 1995, which refers to an economy that is based on digital computing technologies (Bhambri et al., Cloud & IoT, 2022).

In other words, the digital economy is the sum total of economic activity that is a result of billions of everyday online connections or transactions being conducted among people, devices, institutions, businesses, data, and processes.

The development of digital economy has opened a whole new world of possibilities for businesses to grow and contribute to the nation's economy more actively. This new paradigm has allowed the firms to do their business differently as well as more efficiently and cost-effectively.

6.4 CURRENT MARKET SCENARIO

Although cryptocurrency and blockchain are in their initial years as compared to older technologies, some countries like Switzerland are doing exceptionally well in developing and contributing toward blockchain technology. They have recognized the true potential digital currency withholds and even made an initiative toward the betterment of its economic and public sector.

Switzerland not only made cryptocurrency and digital currency a legal asset, but it also collaborated with a French company to provide an integrated service to merchants to accept payments in terms of cryptocurrencies like bitcoin and Ether. More than 100 bitcoin ATMs were also installed in Switzerland to ease the access of bitcoin to its public. Moreover, Switzerland is now focusing on many mature projects like integrating blockchain in the e-Voting system and other public sectors. Two Swiss start-up companies were granted a full Swiss banking license by FINMA (Blockchain, 2021), which sets an example that blockchain integration is the future with Switzerland being one of its torchbearers.

Sweden emerged as one of the first countries globally to embrace the use and grab the opportunities offered by blockchain. The country is currently researching about ways to integrate blockchain in its land administration system (LAS) (Proskurovska et al., 2018).

Many companies and organizations are already working in Sweden, to help Sweden adopt blockchain in different sectors of the economy such as: KnCMiner focuses on producing and supporting miners for cryptocurrencies, and Vorto Gaming is the hub for blockchain-enabled play-to-earn games.

The Swedish Blockchain Association is a non-profit organization; they have the vision to make Sweden a blockchain-ready country. They promoted education, resources, and opportunities in the same field, to meet their vision. Their mission is to engage, educate and empower the community in Sweden (Swedish Blockchain Association, 2022).

Brock Pierce, an American entrepreneur and philanthropist, once said "Blockchain technology is going to change everything more than the internet has," and the recent case of Nigeria justifies this statement to its core, as Nigeria became the first country in Africa to adopt blockchain technology and embed it into its financial system.

The President of Nigeria introduced a digital currency called "eNaira" which is expected to boost cross-border trade, making the transactions more efficient and also improving the monetary policy.

According to the president of Nigeria, "This change can increase Nigeria's gross domestic product by $29 billion over the next 10 years." The claimed amount is significant and can therefore result in the improvement of many sectors along with a hold of good position in terms of the world ranking (Nigeria Issues eNaira, 2022).

Central Bank of Bahama announced the Sand Dollar, with the vision to modernize as well as streamline the financial ecosystem of the country. The sand dollar is expected to decrease service delivery costs, improve transactional efficiency, and increase financial inclusion (Digital Currency, 2022).

Other countries, which embraced the potential blockchain technology possesses and adopted the same for improving their traditional system for better, are mentioned in Table 6.2.

Since every country deals with different economical and geopolitical factors, countries such as South Korea, Bangladesh, China, and Egypt opted for not adopting blockchain technology at present (China Bitcoin, 2022). Table 6.3 refers to their specific reasons for not adopting.

Cryptocurrency has been actively integrating itself into various domains. One such prominent area is digital art or CryptoArt, which is a mixture of the digital art form and blockchain technology. Since 2014 this form is steadily setting its foot in the digital art market and came to be known as a non-fungible token.

TABLE 6.2
Countries that Adopted Cryptocurrency

Country	Mode of adoption	Overview of adoption
Denmark	Adopted as one of the payment methods	People are subject to pay taxes on gains of cryptocurrency, yet it is not completely regulated by the government (Denmark and Cryptocurrency, 2022)
Germany	Inclusion of crypto exchanges in financial services	No specific regulation is enforced but the license is required for operating in crypto-related operations, such as crypto exchange (The Law Reviews, 2022)
France	Legalizes crypto mining	Government enforcement few regulations regarding taxation, ICO, Digital asset services along with mandatory registrations (GLI, 2022).
El Salvador	Bitcoin as legal tender	The government introduced Article 7 which stated that *"every economic agent must accept bitcoin as payment when offered to him by whoever acquires a good or service."* The government also took some positive initiatives such as buying 400 bitcoins (₹118 Cr. approx.) before implementing laws to encourage its citizens (El Salvador, 2022).
Japan	Cryptocurrencies as the legal payment method	The Japanese government amended the Payment Services Act in 2017, which defines "crypto-assets" as payment methods. There are no restrictions on owning and investing in cryptocurrencies (FSA, 2021)

TABLE 6.3
Countries that Banned the Use of Cryptocurrency

Country	Reasons for not adopting cryptocurrency
South Korea	Cybercrime syndicates and money laundering, as regulators view cryptocurrency as a risky financial activity among young retail traders.
Bangladesh	Acoording to the Bangladeshi law, any exchange utilizing bitcoin or other forms of virtual money is unlawful. The violators are liable to be prosecuted and can be sentenced to 12 years in jail.
China	The government's main concern is about crypto mining's effect on the environment. Fraud and money laundering also contribute to government hesitation.
Egypt	National security and the economic health of Egypt as The Egyptian Islamic advisory, Dar al-Ifta, believe that cryptocurrencies might be harmful to the country's national security and economic health (Khang et al., IoT & Healthcare, 2021).
Turkey	Lack of regulation and a central authority for the coins, the government believes crypto-assets can have irreparable damage and have high transaction risks.

A non-fungible token or NFT is non-interchangeable distinctive unit of data stored on a public blockchain network. NFTs work on the principle of scarcity and uniqueness, people can buy and sell such NFTs on different market places such as OpenSea.

The year 2018 was very critical in the history of blockchain and the use of cryptocurrencies. Many startups and projects were announced this year, along with the entry of many big venture capitalists such as Digital Currency Group, which gave this industry a much-needed push. During this period many institutional investors started getting involved to support the development in the field of blockchain. Some of the institutional investors in cryptocurrency are mentioned below:

- Barry Silbert: Founder of the Digital Currency Group, which aims to develop a secure financial system. It focuses on supporting blockchain-based companies and had invested in almost 165 such companies till now. The featured company from the portfolio of DCG is Genesis, which is a liquidity provider. Genesis funds the pool for crypto buyers and sellers. DCG also owns Grayscale investment which holds a total of 6,54,600 bitcoins directly or indirectly.
- Micheal Saylor: Cofounded Microstrategy, a business intelligence firm. Saylor firmly believes that cryptocurrency is the next revolunatiary investment. This firm purchased 1,434 bitcoins in December 2021, adding the firm's inventory of 1,22,478 bitcoins.
- Cameron and Tyler Winklevoss: Owner of Gemini exchange and angel investor for several blockchain-based companies. In December 2021, they were holding around 1,00,000 coins worth around 4.8 million dollars. Gemini exchange launched by Cameron and Tyler in 2015 allows the investors to sell, buy and store their digital assets. Apart from bitcoin, the twins invested in Ethereum, and support the same.

- Elon Musk: Cofounded Tesla, a company that engineers next-generation vehicles and clean energy products. Tesla bought 1.5 billion dollars' worth of bitcoin in 2021 and also started to accept bitcoin as a form of payment (Tesla, 2021).
- Michael Novogratz: Partner at Goldman Sachs frequently comments on the prices of bitcoin. He founded Galaxy Digital Holdings; a brokerage firm that offers a variety of digital asset services related to blockchain. GDH bought $62 million worth of NFT based companies. The company is heavily invested in bitcoin, as of 2021 they had rough holdings of $768 million (Michael NFT Investment, 2022).

According to a 2020 UN report, blockchain the technology lying behind cryptocurrencies could greatly benefit countries who are fighting the global climate crisis and thus help to bring about a far more sustainable global economy.

By the year 2025, blockchain is expected to add a business value of over $176 billion. This would further increase to $3.1 trillion by 2030; by the end of which, 30% of the worldwide customer base will be using blockchain as its fundamental technology.

6.5 CASE STUDY: INDIA

The Ministry of Electronics and Information Technology (MeitY), India published a report "National Strategy on Blockchain" in December 2021 with a vision to create a secure online platform using blockchain framework. MeitY believes promoting research and development in blockchain can make India a global leader in the same (National Strategy, 2022).

NITI Aayog, a policy thinking tank for the government of India, recently collaborated with Gujarat Narmada Valley Fertilizers & Chemicals Limited (GNFC) and developed a blockchain based system that can be used for farm fertilizer subsidy management. The application can provide the use of blockchain features such as immutability and transparency for efficient and effective movement of fertilizer across the supply chain. Several transactions including invoices, claims and challans are recorded on this distributed ledger, which reduces time to activate subsidy, thus providing fertilizers to end customers promptly.

IIT Kanpur (IITK), one of the prestigious engineering colleges in India, is also working on blockchain technology. Their primary focus is on developing e-governance solutions. IITK also launched blockchain-based digital courses, to educate students in the field, with renowned teachers (IIT Kanpur, 2022).

Ripple, a US-based currency exchange company, has launched University Blockchain Research Initiative (UBRI) (Ripple, 2022).

Under this initiative, IIIT Hyderabad (IIITH) was chosen as one of the early partners and the setup of a Centre of Excellence in order to promote the research and adoption of blockchain based solutions.

Through this Centre of Excellence, IIITH is working on the research challenges specific to blockchain such as those related to security or exploring the use of game-theoretic techniques, and evolving an entire next-generation of blockchain solutions.

Tamil Nadu and Telangana have also recently released their policy documents aimed toward adopting the blockchain technology (Telangana, 2022). The government of Telangana is also actively promoting various startups that are based on blockchain (Khang et al., IoT & Healthcare, 2021).

Many real-life use cases have been registered under the initiative taken by the Telangana government. Land registry, use of digital certificates, farm insurance are the top use cases registered by different states of India.

The Reserve Bank of India (RBI) is also exploring the use of blockchain in the banking domain. IBM and Mahindra have jointly collaborated to develop solution for supply chain management (RBI, 2022).

The State Bank of India (SBI) is also working commercial banks and financial institutions toward a blockchain-based application pilot. Yes, Bank, Axis Bank, and ICICI Bank are also slowly adopting blockchain in their banking use cases.

The Finance Miniter of India believes that "Introduction of a central bank digital currency will give a big boost to the digital economy; digital currency will also lead to a more efficient and cheaper currency management system."

Recently, the annual budget was presented by the honourable Finance Minister of India, Nirmala Sitaraman in which the following key points were announced regarding the use of cryptocurrency:

- Any income derived from the virtual digital assets is taxable at the rate of 30%.
- There will be no deduction with exception of the cost of acquisition.
- The TDS is applicable beyond a specified monetary threshold.
- The gift of virtual currencies is taxable in the hands of the recipient.
- The country's central bank, the Reserve Bank of India (RBI), will introduce a digital currency in the next financial year using blockchain and other supporting technology.

Up until now, the Indian government didn't have a stand in the matter of cryptocurrencies and blockchain, but now they've given a warm welcome to this booming technology. The government has opened its gates to blockchain-based companies, but this announcement could be a two-faced decision.

On one hand, the government is legalizing cryptocurrencies directly or indirectly but on the other hand, they imposed a heavy tax on the same. Such heavy taxes wouldn't be a feasible option for the common citizen to invest or to trade in the same. These taxes and TDS will restrict a middle-class citizen, which is in majority, to actively take part in this revolutionary technology.

For attracting this sector government has to take some serious measures such as relaxation in tax slab, spread awareness about this technology and run programs to educate people in the field of blockchain, giving some relaxation to blockchain-based companies to let them evolve in this competitive environment.

6.6 APPLICATION AREAS

Blockchain can be implemented across many different sectors and services, such as in financial services, for managing decentralized portfolio; in the governance sector, for minimizing the role of the middle man; in the health sector, for developing decentralized health records of patients for all across the country and other sectors. These applications are discussed as follows:

6.6.1 FINANCIAL SERVICES

Blockchain simplifies asset management, all the processes, and actions in managing assets, i.e. portfolio management, trading/transaction of cryptocurrency (Cryptocurrency in India, 2022). Record encryption also reduces errors in the process. Blockchain made money transfer and online payments more secure, fast, and less prone to delays and errors.

The backend work is done on a smart contract, which is automated in such a way that the contract will be executed once the required conditions are met. The global value of cross-border payments in the retail sector was predicted to rise from $1.95 trillion in the year 2016 to $3.56 trillion by 2022.

When such an immense amount is concerned, security, time consumption, and a smaller number of errors are a massive responsibility and blockchain is a life-saving technology as it deduces the majority of the flaws of a centralized network. For instance, the encryption features of blockchain oversee any fraudulent or suspicious actions in an account.

WazirX is an Indian crypto exchange that ranked among the top 100 secured crypto exchanges globally. Its noteworthy feature is that it stores 95% of its total funds offline, which keeps it safe from hacking activities (WazirX, 2022).

6.6.2 GOVERNMENT SECTOR

Corruption on which every government spends millions of dollars to eradicate, occurs due to the presence of middle parties. Various governments are already experimenting different solution for blockchain-based land title registries such as Sweden. Smart contract-based land registries can potentially minimize the opportunity for self-interested manipulation of land rights and thereby increase the resilience of land ownership more generally.

6.6.3 HEALTH SECTOR

Blockchain could facilitate the safe and secure transfer of the patient medical history, and manage the entire medical medicine supply chain. Blockchain's security safeguards the digital identity of any individual or patient with complex and secure codes that can protect the sensitive nature of medical data. The decentralized nature of the technology also permits patients, doctors, and healthcare providers to share the same information effectively and efficiently.

6.6.4 ENTERTAINMENT SECTOR

The ledger provides transparent transmission of royalties to everyone included in the label. It also helps musicians and other artists to quickly monitor and claim royalties for their work.

One such real-life application of this sector is NFTs, which are digital assets minted through smart contracts. NFTs are based on the concept of uniqueness, and scarcity, as they are created in limited quantity and can be duplicated at a cost by those who own them (NFTs, 2022).

6.6.5 DIGITAL IDENTIFICATION

Companies and financial institutions are strictly ordered to follow Know-Your-Customer (KYC) initiatives, but many providers lack resources and are unable to sufficiently meet the demands of the government.

Blockchain can reduce the need for in-person identification, and replace it with Digital IDs. This change in the method can protect people from identity theft, and allow millions to control their identity digitally. Blockchain-based identification can ensure that no third party can access your data without consent (Budget India, 2022).

Many blockchain-based companies have emerged for providing various services in different sectors such as Finance, Art, Music, Healthcare; some are mentioned below:

- Akiri specializes in operating in the healthcare industry. Akiri's mission is to protect patient data while transferring. Its differentiated feature is that it doesn't store data in its system while operating as a protocol to set policies. Its system configures data in real-time while checking the source and destination of data. To make the healthcare system more secure Akiri ensures that data is sharable with authorized parties only.
- Brave browser is an open-source web browser which was developed by Brave.
- Software. Its unique feature is that it automatically blocks online advertisements and trackers embedded in websites. A reward is also one of the key features of the Brave browser, in which the user gets paid in BAT (cryptocurrency) if he opts for viewing the advertisement.
- Rarible platform is a digital art marketplace based on Ethereum, that facilitates the sale, and purchase of digital art works. It uses two NFT token standards, ERC-721, and ERC-1155 (Rarible NFT Marketplace, 2022).
- Aave is a one of the biggest DeFi platforms that enables the borrowing and lending of cryptocurrency (DeFi, 2022). It including so-called "flash loans," which are considered to be the first uncollateralized loan option in the DeFi space. Aave is developed on the Ethereum protocol. Lenders can earn relatively low-risk, passive income from interest paid on loans, without getting involved with third parties.
- Token exchange: Uniswap is a automated liquidity provider which make it easy to exchange Ethereum tokens. There is no orderbook or central facilitator on

Uniswap, instead tokens are exchanged through liquidity pools that are defined by smart contracts.
- Covantis is a blockchain-based solution developed by ConsenSys. Its primary objective is to provide an efficient network for the execution of a huge amount of agricultural trade (Covantis go-live, 2022).
- MetaMask Swaps is a combined work of MetaMask and Codefi. It focuses on providing its user the lowest network fee in the market, by collecting information from different exchanges and comparing them (Blockchain Case Study, 2022).
- Mata Capital operates in the field of real estate management. They partnered with Codefi to incorporate blockchain technology for bringing more efficiency and transparency to the system.
- HMLR is a digital asset management platform that customized UK's land registry in under 3 weeks (HM Land Registry, 2022).
- Komgo provides a platform for commodity trade, with the help of blockchain. They are one of the world's leading commodity trade and finance companies (Komgo, 2022).
- August Debouzy is a law firm that utilizes blockchain to offer its clients a Security Token Offering service (Debouzy, 2022).
- Project i2i focuses on the betterment of the rural banking sector by providing them a platform with better efficiency. The i2i is based on the Ethereum network and has a partnership with Unionbankof Philippines (Project i2i, 2022).
- Project Khokha provides the solution for real-time gross settlement (Khokha, 2022). Its platform is based on blockchain technology and was built by South Africa Reserve Bank (Reserve Bank, 2018).
- Project Ubin (Project Ubin, 2022) is an effort by Singapore Monetary Authority to strengthen institutional infrastructure by providing gross settlement with security and privacy (Singapore Monetary Authority, 2022).

6.7 CHALLENGES

New technology comes with new challenges, so does blockchain. Although theoretically, this technology can solve a tremendous number of problems in our current framework, blockchain has its challenges too, which are related to scalability, resources, and jurisdictions. These are discussed in length below:

- Jurisdiction dilemma: As decentralized applications are distributed, nodes can be situated in different parts of countries and thus different jurisdictions will impose. This makes it very challenging to determine which rules, policies, or laws are functional.
- Lack of framework: Blockchain applications introduce new forms of decentral value generation. A compliance framework is, commonly accepted decentralized governance rules, risk management practices, and compliance frameworks are necessary.

- Scalability: Since every node stores replica of every transaction that takes place on the blockchain. Scalability becomes is one of the major concerns when an ample amount of data is stored on daily basis for a very long time
- Safety concern: Due to lack of proper framework, no safety standards are established. So the risk of fraud and cyberattacks is always present while using blockchain.
- Transaction speed: Bitcoin blockchain can only process seven transactions a second, while Alibaba e-commerce platform, processes 325 transactions per second. There is still a huge gap between no transactions which needs to improve to a great extent, before using blockchain for commercial use.
- Energy consumption: Proof of Work is the method by which blockchain stores data. It consumes 66.7 terawatt-hours per year, which is so huge, it can be compared to the total energy consumed by the Czech Republic, a country of 10.6 million people.

Blockchain is a promising technology, and when specifically used with other technologies it offers organisations a space to re-establish their internal and external processes. It remove inefficiencies, increase transparency and make any organisation more efficient. It faces several challenges that could affect its adoption across countries, but a significant more research effort could help overcome the considerable governance, technological and environmental anomalies involved therein (Budget, 2022).

6.8 CONCLUSION

Blockchain has become one of the revolutionary technologies over the last decade. The most notable application of blockchain is in today's financial sector and cryptocurrency is the face for that application. In this work, we analyze the adoption of blockchain on a global scale and we discussed the challenges in the adoption of blockchain.

We presented a case study on India and various applications of blockchain are also presented with a special focus on business use cases of blockchain. Cryptocurrencies are here to stay and will surely play a significant role in boosting the future of the digital economy.

We believe this work will greatly help in reviewing and taking a critical stance on the subject.

REFERENCES

About – Hyperledger Foundation. www.hyperledger.org/about (accessed Feb. 07, 2022).

Azure Blockchain Workbench Preview overview – Azure Blockchain | Microsoft Docs. https://docs.microsoft.com/en-us/azure/blockchain/workbench/overview (accessed Feb. 7, 2022).

B. A. V., B. D. D., P. W. W., & T. A. S., Cryptocurrency and blockchain technology in the digital economy: development genesis, π-Economy. 67(5), 9–22, 2017, doi: 10.18721/ JE.10501.

Beck, R., European Blockchain Center, Fraunhofer-Institut für Arbeitswirtschaft und Organisation, & Industriens Fond, Study on the economic impact of blockchain on the Danish industry and labour market. European Blockchain Center, 2019.

Bhambri, Pankaj, Sita Rani, Gaurav Gupta, & Khang, A., *Cloud and Fog Computing Platforms for Internet of Things*. Chapman & Hall. ISBN: 9781032101507, 2022, doi: 10.1201, www.routledge.com/Cloud-and-Fog-Computing-Platforms-for-Internet-of-Things/Bhambri-Rani-Gupta-Khang/p/book/9781032101507

Bitcoin_Whitepaper_Document_HD.

Blockchain 2021 – Technology – Switzerland. www.mondaq.com/fin-tech/1127348/blockchain-2021 (accessed Dec. 11, 2021).

Blockchain & Cryptocurrency Laws and Regulations | France | GLI, Accessed: Jan. 15, 2022. [Online]. Available: www.globallegalinsights.com/practice-areas/blockchain-laws-and-regulations/france

Blockchain Case Study | Mata Capital: Real Estate Investments with Blockchain | Codefi Assets | ConsenSys. https://consensys.net/codefi/assets/mata-capital/ (accessed Feb. 8, 2022).

Blockchain Case Study | MetaMask Swaps: Removing User Frictions from P2P Trading | ConsenSys. https://consensys.net/codefi/markets/metamask-swaps/ (accessed Feb. 8, 2022).

Budget 2022 | Here's what Finance Minister Nirmala Sitharaman said about taxing cryptocurrencies. www.moneycontrol.com/news/business/cryptocurrency/budget-2022-heres-what-finance-minister-nirmala-sitharaman-said-about-taxing-cryptocurrencies-8019751.html (accessed Feb. 8, 2022).

Budget India: Find all results about Budget India including the latest news, photos, videos | Hindustan Times. www.hindustantimes.com/business/crypto-tax-is-here-india-imposes-30-tax-on-proceeds-of-digital-assets-101643698281541.html (accessed Feb. 5, 2022).

Business Insider India, 2020. What caused the Great Recession? Understanding the key factors that led to one of the worst economic downturns in US history, *Business Insider India*. www.businessinsider.in/finance/news/what-caused-the-great-recession-understanding-the-key-factors-that-led-to-one-of-the-worst-economic-downturns-in-us-history/articleshow/84253779.cms (accessed Feb. 7, 2022).

Catalini, C. 2022. How blockchain technology will impact the digital economy the platform of the future? https://ide.mit.edu/sites/default/files/publications/IDE%20Research%20Paper_v0517.pdf

Chakrabarti, A., & A. K. Chaudhuri, Blockchain and its scope in retail, *International Research Journal of Engineering and Technology*, 2017, [Online]. Available: www.irjet.net

China Bitcoin, 2022. China already banned Bitcoin mining – now it's cracking down on holdouts, *Fortune*. https://fortune.com/2021/11/17/china-bitcoin-mining-ban-crypto-holdouts-ether-solana-price/ (accessed Jan. 9, 2022).

Combating Corruption, 2022. www.worldbank.org/en/topic/governance/brief/anti-corruption (accessed Feb. 8, 2022).

Covantis go-live, 2022. *ANEC*. https://anec.com.br/article/covantis-go-live (accessed Feb. 8, 2022).

Cryptocurrency Bill, 2022. These are the countries where cryptocurrency is restricted or illegal, *BusinessToday*. www.businesstoday.in/crypto/story/cryptocurrency-bill-these-are-the-countries-where-cryptocurrency-is-restricted-or-illegal-313247-2021-11-24 (accessed Jan. 9, 2022).

Cryptocurrency in India, 2022. The past, present and uncertain future, *The Economic Times*. https://economictimes.indiatimes.com/tech/trendspotting/cryptocurrency-in-india-the-past-present-and-uncertain-future/articleshow/81410792.cms?from=mdr (accessed Feb. 19, 2022).

Debouzy, 2022. Cabinet d'avocats – August Debouzy. www.august-debouzy.com/en/ (accessed Feb. 8, 2022).

DeFi, 2022. Understanding DeFi and its importance in the crypto economy. www.coindesk.com/tech/2022/01/20/understanding-defi-and-its-importance-in-the-crypto-economy/ (accessed Feb. 19, 2022).

Denmark and Cryptocurrency – Virtual Currency Laws, *Freeman Law*. https://freemanlaw.com/cryptocurrency/denmark/ (accessed Jan. 15, 2022).

Deshmukh, P., G. Kulkarni, M. Shaikh, V. Taral, & K. S. Thakare, Digital India digital economy using BCT, doi: 10.51319/2456-0774.2021.6.0079.

Digital Currency, 2022. The Sand Dollar is the world's most advanced "official" digital currency. www.consulting.us/news/6047/the-sand-dollar-is-the-worlds-most-advanced-official-digital-currency (accessed Feb. 7, 2022).

Digital Economy Compass 2020.

El Salvador: Bitcoin Law Regulation, *Consortium Legal Eng*. https://consortiumlegal.com/en/el-salvador-bitcoin-law-regulation/ (accessed Jan. 15, 2022).

Forbes India, Don't ban bitcoin. It's good for the economy. www.forbesindia.com/article/take-one-big-story-of-the-day/dont-ban-bitcoin-its-good-for-the-economy/66335/1 (accessed Jan. 16, 2022).

Free Crypto Screener, *BitScreener*. https://bitscreener.com/ (accessed Jan. 23, 2022).

Global spending on blockchain solutions 2024, *Statista*. www.statista.com/statistics/800426/worldwide-blockchain-solutions-spending/#statisticContainer (accessed Dec. 10, 2021).

Hassani, H., X. Huang, & E. Silva, Banking with blockchain-ed big data. https://doi.org/10.1080/23270012.2018.1528900, 5(4), 256–75, Oct. 2018, doi: 10.1080/23270012.2018.1528900

HM Land Registry, 2022. www.gov.uk/government/organisations/land-registry (accessed Feb. 8, 2022).

Humphrey, C. Barter and Economic Disintegration, *Man* 20(1), 52, Mar. 1985, doi: 10.2307/2802221

IIT Kanpur, 2022. IIT Kanpur launches blockchain-based digital degrees; here's what it means, *TechGig*. https://content.techgig.com/iit-kanpur-launches-blockchain-based-digital-degree/articleshow/88564128.cms (accessed Feb. 8, 2022).

Khang, A., Geeta Rana, Ravindra Sharma, Alok Kumar Goel, & Ashok Kumar Dubey, The Role of Artificial Intelligence in Blockchain Applications, *Reinventing Manufacturing and Business Processes through Artificial Intelligence*, 2021, doi: 10.1201/9781003145011

Khang, A., Rani, S., & Sivaraman, A.K. (Eds.). (2022). *AI-Centric Smart City Ecosystems: Technologies, Design and Implementation* (1st ed.). CRC Press. https://doi.org/10.1201/9781003252542

Khokha, 2022. Blockchain for Central Banks: Project Khokha Case Study | ConsenSys. https://consensys.net/blockchain-use-cases/finance/project-khokha/ (accessed Feb. 8, 2022).

Komgo, 2022. About Komgo. www.komgo.io/about (accessed Feb. 8, 2022).

Kumar, G. et al., Decentralized accessibility of e-commerce products through blockchain technology, *Sustainable Cities and Society* 62, 102361, Nov. 2020, doi: 10.1016/J.SCS.2020.102361.

The Law Reviews, *The Virtual Currency Regulation Review*. https://thelawreviews.co.uk/title/the-virtual-currency-regulation-review/germany (accessed Jan. 15, 2022).

Maxmudjanovna, A. I., P. Rano Abdurasulovna, & N. N. Erkinovna, The future of the digital economy: Concept and role of blockchain technologies, *Journal of Critical Reviews*, 2020, doi: 10.31838/jcr.07.08.350

Michael NFT Investment, 2022. Michael Jordan's NFT Investment, *Unchained Podcast.* https://unchainedpodcast.com/michael-jordans-nft-investment/ (accessed Feb. 7, 2022).

National Strategy, 2022. National Strategy on Blockchain towards Enabling Trusted Digital Platforms.

News Release: DHS awards $197k for tracking raw material imports, *Homeland Security.* www.dhs.gov/science-and-technology/news/2019/11/18/news-release-dhs-awards-197k-tracking-raw-material-imports (accessed Dec. 13, 2021).

NFTs, 2022. 15 Most Expensive NFTs Sold (So Far) | ScreenRant. https://screenrant.com/expensive-nfts-sold-so-far/ (accessed Jan. 9, 2022).

Nigeria Issues eNaira (NGN USD) Digital Currency after Banning Crypto Exchange, *Bloomberg.* www.bloomberg.com/news/articles/2021-10-25/nigeria-starts-digital-currency-after-banning-crypto-exchange (accessed Jan. 8, 2022).

Pilkington, M. *Blockchain Technology: Principles and Applications*, Research Handbooks on Digital Transformations, pp. 225–53, Sep. 2016, doi: 10.4337/9781784717766.00019.

Project i2i, 2022. Blockchain Payments: Project i2i Case Study | ConsenSys. https://consensys.net/blockchain-use-cases/finance/project-i2i/ (accessed Feb. 8, 2022).

Project Ubin, 2022. Central Bank Digital Money using Distributed Ledger Technology. www.mas.gov.sg/schemes-and-initiatives/project-ubin (accessed Feb. 8, 2022).

Proskurovska, A., & S. Dörry, Is a Blockchain-Based Conveyance System the Next Step in the Financialisation of Housing?: The Case of Sweden, *SSRN Electronic Journal*, Nov. 2018, doi: 10.2139/ssrn.3267138.

Queiroz, M. M., R. Telles, & S. H. Bonilla, Blockchain and supply chain management integration: a systematic review of the literature, *Supply Chain Management* 25(2), 241–54, Feb. 2020, doi: 10.1108/SCM-03-2018-0143/FULL/XML.

Rani, Sita, Khang, A., Meetali Chauhan, & Aman Kataria, IoT equipped intelligent distributed framework for smart healthcare systems, *Networking and Internet Architecture*, 2021, https://arxiv.org/abs/2110.04997v2, doi: 10.48550/arXiv.2110.04997.

Rarible NFT Marketplace, 2022. RARI Coin & NFT Crypto Art, *Gemini.* www.gemini.com/cryptopedia/rarible-nft-art-marketplace-rari-coin#section-rarible-features-nft-royalties-minting-and-more (accessed Feb. 4, 2022).

RBI, 2022. RBI ban on cryptocurrency: Anger, shock, confusion as RBI bars banks from cryptocurrencies, *The Economic Times.* https://economictimes.indiatimes.com/markets/stocks/news/anger-shock-confusion-as-rbi-bars-banks-from-cryptocurrencies/articleshow/63638799.cms (accessed Feb. 19, 2022).

Reserve Bank, 2018. Prohibition on dealing in Virtual Currencies (VCs) Reserve Bank has repeatedly through its public notices on, 2018.

Ripple, 2022. Ripple, IIITH Blockchain Center of Excellence – Machine Learning Lab. https://mll.iiit.ac.in/ripple-iiith-blockchain-coe/ (accessed Feb. 8, 2022).

Singapore Monetary Authority, 2022. Blockchain Market Size, Share and Global Market Forecast to 2026 | Markets and Markets. www.marketsandmarkets.com/Market-Reports/blockchain-technology-market-90100890.html (accessed Feb. 3, 2022).

Statista, 2021. Value of global cross-border payments by type 2022, *Statista.* www.statista.com/statistics/609723/value-of-cross-border-payments-by-type/ (accessed Dec. 13, 2021).

Swedish Blockchain Association. https://swedishblockchain.se/ (accessed Feb. 7, 2022).

Swinburne News, 2016. The DAO: a radical experiment that could be the future of decentralised governance, *Swinburne News.* https://web.archive.org/web/20160516134246/www.swinburne.edu.au/news/latest-news/2016/05/the-radical-dao-experiment.php (accessed Jan. 3, 2022).

Telangana, 2022. Telangana calls entrepreneurs to pilot blockchain solutions in state, *Hyderabad News – Times of India*. https://timesofindia.indiatimes.com/city/hyderabad/t-calls-entrepreneurs-to-pilot-blockchain-solutions-in-state/articleshow/88348334.cms (accessed Feb. 8, 2022).

Tesla, 2021. Tesla held nearly $2 billion worth of bitcoin at the end of 2021. www.cnbc.com/2022/02/07/tesla-held-nearly-2-billion-worth-of-bitcoin-at-the-end-of-2021.html (accessed Feb. 7, 2022).

Tomar, R. S. Maintaining trust in VANETs using blockchain, *ACM SIGAda Ada Letters* 40(1), 91–6, 2020.

Travel Rule Crypto in Japan by FSA [2021], *Notabene*. https://notabene.id/world/japan (accessed Jan. 15, 2022).

WazirX, 2022. Is WazirX safe?, *Javatpoint*. www.javatpoint.com/is-wazirx-safe (accessed Jan. 8, 2022).

What is Aave? How to buy AAVE Crypto, *SoFi*. www.sofi.com/learn/content/what-is-aave-crypto/ (accessed Feb. 4, 2022).

What is blockchain technology blockchain technology? How does it work?, *Built-In*. https://builtin.com/blockchain (accessed Jan. 8, 2022).

What is decentralization? https://aws.amazon.com/blockchain/decentralization-in-blockchain/ (accessed Jan. 8, 2022).

Zeilinger, M. Digital art as "monetised graphics": Enforcing intellectual property on the blockchain, *Philosophy & Technology 2016* 31(1), 15–41, Nov. 2016, doi: 10.1007/S13347-016-0243-1

7 Application of Blockchain in Online Learning
Findings in Higher Education Certification

Pushan Kumar Datta and Susanta Mitra

CONTENTS

7.1 Introduction 103
7.2 Problem Statement 105
7.3 Perspective and Challenges 106
    7.3.1 Intervention for Stabilization of Supply and Demand......................106
    7.3.2 Facilitate 106
    7.3.3 Assistance..106
    7.3.4 Record 107
    7.3.5 Trusted Certification 107
    7.3.6 Decentralized Sharing 107
    7.3.7 Provide Income Support 108
7.4 Intensified Cooperation with Competent World Organizations 108
7.5 Conclusion..109
References..109

7.1 INTRODUCTION

The public's interest in blockchain technology has grown because of its many benefits, including security, quicker transactions, and lower prices, anonymity, and data integrity.

Bitcoin's success has prompted the use of blockchain technology in a variety of fields, including the financial market, IoT, supply chain, voting, medical treatment, and storage (Bhambri et al., Cloud & IoT, 2022). Smart contracts, which represent automated execution of transactions where terms of the transactions are embedded (Chen et al., 2018) in computer code that is automatically executed by the programme upon recognition of a certain input, are credited with faster transactions.

Furthermore, the student desires more rapid engagement than is often achievable in such networks. Even if they do not lead to subsequent conversation, occasional intensive encounters when solving a problem or working hard to comprehend something may be highly developing and fulfilling (Gräther et al., 2018). As a consequence, the United Nations' fourth Sustainable Development Goal, ensuring inclusive and quality education and promoting access to higher education for everybody, is highlighted in this section.

As a result, countries all over the world strive to provide good education, which is characterized by variables such as teacher skills, student attendance, student-to-teacher ratio, government education spending, and test results. Learners, learning settings, content, process, and results all play a role in providing high-quality education through virtual learning. Despite their popularization, present virtual learning forms and processes have several flaws in the face of an increasingly open and digital Web.

For example, MOOCs' learning processes lack recognition and official certification; students' privacy is jeopardized because courses and data security are solely dependent on a centralized online education platform; students' intellectual property cannot be effectively protected due to the openness of the Internet and data privacy concerns; and there is no open ended cross-platform course sharing mechanism to fully share teaching resources (Rivera-Vargas et al., 2019).

It is necessary to develop a distributed and trustable data storage mechanism to record the students' learning process to make the learning process and outcomes trustworthy. Teachers must master instructional methods and be able to conduct thorough student evaluation and assessment to ensure effectiveness.

A teacher wants to contact another instructor who is teaching a comparable course on the same subject because he wants to create a work task for his pupils or just trade lectures to broaden viewpoints. Smart contracts, ranging from securities, stocks to bank loans, can be enforced automatically without human intervention (Kosba et al., 2016).

In terms of the IoT, blockchain technology enables devices to communicate autonomously and identify errors (Emilie et al., 2016). This technology has also been applied preliminarily in the field of education. For instance, Mike Sharple proposed that blockchain can be employed to realize distributed storage of education data, forming the so-called knowledge currency (Sharples et al., 2016).

Some scholars suggested applying blockchain in credit card authentication, secure data encryption and distributed data storage (Sun et al., 2018). The MIT Me-dia Lab built a digital learning certificate system using blockchain technology and Mozilla's open badge (Redman et al., 2018). In addition, blockchain technology has been adopted for product design in the industrial field.

Sony Global Education (Global, S. G. S, 2016), a blockchain technology infrastructure platform under Sony Corporation of Japan, can openly and securely share learning courses and record data, without disclosing this information to the education management authority, thus realizing the fairness and digitization of education.

University College London uses blockchain technology to help postgraduates of financial risk management verify the authenticity of their academic qualifications (Williams et al., 2019). Online education, often known as remote schooling or digital training, is a web-based teaching approach that uses information technology and the Internet to disseminate knowledge and facilitate learning.

With the Internet as the medium, online teaching breaks down barriers such as location, environment, time, and teachers, and provides students with high-quality learning opportunities at any time and from any location. A micro-credential is a certificate that demonstrates learning results from a brief, publicly assessed course or module.

Micro credentials can be earned in person, online, or in a hybrid mode (Yang et al., 2017). Because of the flexibility of these certifications, individuals, especially those in full-time job, may take advantage of learning possibilities. Micro-credentials are therefore a very flexible and inclusive kind of learning that allows for the focused development of skills and competencies.

Higher education and vocational education and training (VET) institutions, as well as private organizations, provide micro-credentials (Cheng et al., 2018). Vest's focus is on open education, open knowledge, and making science accessible to anybody who is interested. Vest mentions the creation of the meta-university, which he defines as "a transcending, accessible, powerful, dynamic, communally produced framework of open resources and technologies on which much of higher education at all levels might be developed or enhanced" in the same chapter (Kamišalić et al., 2019).

7.2 PROBLEM STATEMENT

The advantages of the proposed VET system include: linking blockchain and higher education diplomas; less diploma counterfeiting and fraud (due to decentralized management); time and money savings – especially for the younger generations; and increased meritocracy in academia and the job market (as real qualifications are accessed for processing by higher education institutions and firms).

Enhanced decentralization and data quality (accurate, verified, and validated data) using blockchain and smart contracts (Duan et al., 2017); increased data security decreases the danger of fraud; enhanced decentralization and data quality (correct, verified, and validated data). However, for the notion to become popular, it must be tested; standards and implementation obstacles remain. Higher education institutions, as well as their courses and degrees, are accredited and included in a variety of national and worldwide rankings (Gresch et al., 2019).

Employers and higher education institutions will want to know: what qualifications applicants have; and how good those qualifications are (relative to other applicants and institutions which concede academic certificates). Additionally, ethical considerations may be incorporated (Vidal et al., 2019) into the process, since this is becoming a growing concern in society and the educational sector, especially in undergraduate medical ethics education and other domains where the humane component is essential.

The advent of smart contracts marks the start of Blockchain 2.0, with the creation of a new set of financial applications. Many solutions are being created because of the increased interest of numerous different organizations and sectors, mostly due to blockchain's core properties of decentralization, immutability, and transparency. As a result, we have entered the Blockchain 3.0 phase.

Higher education is one of the businesses that might profit greatly from the new technology, as the requirement for document authenticity, transparency, and trust intersect with blockchain features in a perfect match. To build the blockchain, the blocks are cryptographed and connected. Blockchain technology, with its proven results and security, is an effective response to the challenges of online education.

Specifically, blockchain could provide full and non-temperable training records for online learning without the need for third-party monitoring, as well as ensure impartial course credit certification. Moreover, smart contracts can increase the efficiency of course distribution in online education, whereas cryptographic-based data processing guarantees that users' privacy is protected. As a result, utilizing blockchain technology in online education is a worthwhile initiative.

7.3 PERSPECTIVE AND CHALLENGES

7.3.1 Intervention for Stabilization of Supply and Demand

The covid pandemic created a crisis suddenly that brought in immediate and massive market disorder. In a situation when the invisible hand understandably is imperfect to stabilize the consequences of externally introduced disturbances by spontaneous adjustments of supply and demand, the intervention of governments is not only justifiable but necessary.

As never before, the approach of state intervention is questioned by no one (San et al., 2019). As already stated, the education ministries should perform constant monitoring on changes of independent variables in human resource development for the key education outcomes and apply appropriate corrective actions to mitigate the disturbance.

7.3.2 Facilitate

Facilitate the availability of educated labor in pandemic for agricultural sector for creating jobs. Agricultural production as a highly labor-intensive activity can be easily threatened by the restrictions on movement of people. A possible scenario is that epidemiological outbreak may affect family farms that are dominantly operated by elderly farmers as the most vulnerable category.

To facilitate the availability of labor force in the affected sectors, the authorities can offer financial support for partial compensation of the gross salary for employees in the agricultural sector to do some distance learning course.

The payments would assist the producers' financial flow and improve the attractiveness of agricultural jobs. In a situation of the shutdown of businesses in many industries, the agriculture may act as a buffer for the surplus of labor, especially one who previously migrated from the rural areas to higher-paid professions in cities or abroad.

7.3.3 Assistance

Assistance to the businesses in difficulty to run educational institutes. The coronavirus has shocked supply and demand. For different reasons, many family farms and big companies face difficulties in their functioning. Some lost their suppliers and markets, have problems finding seasonal workers or maintain current employees with a reduced volume of operations.

A lot faced with a reduction in cash-flows and revenues as sales fall, while the fixed costs and salaries payments keep running. Carefully monitoring the situation and taking a series of measures can ease the pressure on food businesses. Support offered in the form of know-how and working capital is what might be critical for many educational businesses to cope with the resulting situation.

7.3.4 RECORD

Record of the learning trajectory, data is stored in a distributed database on the blockchain, which uses timestamps to organize data blocks in chronological order. The new data blocks are not able to be removed. The cryptographic technique is used to prevent data alteration, which makes fraud more difficult.

Most online education sites are now decentralized, with courses of varying quality. Worse, owing to the lack of a consistent certification system, the learning outcomes are not publicly recognized. Surprisingly, internet schooling does not have positive benefits. The blockchain's chronological data recording provides an excellent means to capture online education learning data.

7.3.5 TRUSTED CERTIFICATION

Trusted certification of learning records, the blockchain-based learning record not only completely records students' learning data, but it also resists manipulation and removal, providing a solid assurance of the data's reliability. Simultaneously, learning data may be broadcast over the network and simply collected by the employer, whose security is maintained by encryption technology.

The company may discover more about the students' learning status and validate their information using blockchain-based data. As a result, blockchain technology may efficiently prevent paper fraud, counterfeit academic certificates, and other misconduct in higher education, while also establishing a trustworthy platform for students, teachers, and employers.

7.3.6 DECENTRALIZED SHARING

Decentralized sharing of education resources, there are several online education platforms available today that provide a wide range of courses with extensive material. Nonetheless, because to limitations such as instruction mode, copyright, and other factors, the courses are not shareable between platforms.

The user experience for people taking various sorts of courses is unsatisfactory since they must log into many platforms. Similarly, students of higher education find it challenging to study knowledge from another school or field. Because of a lack of uniform and effective exploitation, many high-quality course materials are squandered (Kirilova et al., 2018).

With the advent of the sharing economy (for example, shared bikes), society is demanding for greater resource usage. In the sphere of education, resource sharing is the way things are going to go in the future.

7.3.7 PROVIDE INCOME SUPPORT

The radical deployment of public funds to combat the threat posed by a novel coronavirus can push governments to restrictions of previously planned public finances including the support scheme for education and rural development. While capital investment projects obviously won't be implemented, it is not advisable to downsize the funds allocated for farmers' income support.

Moreover, monitoring the incomes should be permanently conducted, especially where there are indications for a slump of selling prices. Based on the calculations of rural students' financial situation, the authorities could introduce intervention direct payments aimed to offset the loss of incomes.

The emergency financial transfers will not only strengthen endangered production but also serve as a safety net for the rural population, especially for the vulnerable categories. The countries operating programs for support of farmers' security funds can consider additional capitalization for upholding the income assistance to their members.

7.4 INTENSIFIED COOPERATION WITH COMPETENT WORLD ORGANIZATIONS

This crisis is global; thus the crisis management requires co-operation and assistance that cross far beyond the national borders. To exchange experiences and know-how, as well as if needed to set request for support (such as the supply of agricultural inputs, technical assistance for policy formulation, etc), the ministries should further strengthen their contacts with the international global institutions dealing with education etc.

Similarly, the international aid agencies should be communicated for possible bilateral assistance and reprogramming of previously stated objectives in the light of overcoming the Covid19 consequences. India has seen the greatest movement of workers in the first two months of lockdown.

Based on the following studies, it can be found that virtual ICTs are "enveloping our reality" in the sense that we utilize them for complex or time-consuming activities and then allow them to perform their work in their designated space (called envelope) in a machine-like manner.

ICTs are also "re-ontologizing our reality," (Bucea-Manea-Țoniș et al., 2021) as certain terms take on new meanings and connotations in different contexts. "Presence," for example, has lost some of its implications of physical closeness and now simply means "being intractable." Online education has become a new option for individuals to obtain information because of the advancement of Internet technology (Ocheja et al., 2018).

However, there are still issues with online education, such as the lack of outcomes certification, insufficient privacy, and the lack of a sharing mechanism. Due to its decentralized, de-trusted, and dependable characteristics, blockchain has been widely accepted as an emerging computer technology in a variety of areas (Williams et al., 2019).

As a result, this chapter blends blockchain technology with online education (Serranito et al., 2020) to address the issues, resulting in a smart, decentralized

(Nikolskaia et al., 2019), and collaborative online education system (Capece et al., 2020). The outcomes of the study indicate that online education is on the rise.

7.5 CONCLUSION

The following discussion sets are the advantages of blockchain in online education, it can automate university administration and focus on personal traits in human contact by using the power of ICTs, notably AI (Mikroyannidis et al., 2018) instead of using recorded lectures to industrialize instruction.

It may drop all the outdated and misleading "distance-," "e-," "online-," and other learning labels and focus (Serranito et al., 2020) just on teaching and learning communication, regardless of geography, technology, or other factors. It appears to be completely compatible with the use of learning analytics, adaptive learning, and other modern ICTs to assist teachers in determining how to assist, comprehend, and correct students (Khang et al., IoT & Healthcare, 2021). It acknowledges that delivering lessons is not the most essential point a scholar can do.

More importance is put on personal supervision, feedback on students' progress, and critical discussions. It works well with the time-based blended learning concept that attempts to create a process of learning. It has been criticized for its links to old exclusive classic education, when a few gentlemen students just convened for talks in scholars' quarters (Khang et al., AI & Blockchain, 2021).

I'm sure it's good, but is it scalable to the masses? In these each scenario, how else is quality control handled? Is blockchain truly required for this concept, aside from keeping exam records secure? It will take a very long time for such a concept to be adopted and understood. It is disruptive, and it will cause long-term problems with existing structures.

Unbundling schooling would benefit the notion even more: to distinguish between examination and instruction. But what does this mean in terms of administration? The disadvantage of the work is that we can see a study is functioning for (a) primarily text/content-based courses; (b) mature people who want to study rather than just spend time on campus and get a degree along the way; and (c) experienced academics who are engaged in student learning and have good communication skills (Khang et al., AI-Centric Smart City Ecosystem, 2022).

However, there is a lot in today's higher education that does not fit within this cross-section. Higher education and its "Maecenas" might see this as a risky way to offer good PG certificates or master's degrees without a central control system, because it means that normal education (universities and institutes) will not accept those credentials. In this type of institution, the professors may be the new "guilds" or "lodges" of knowledge.

REFERENCES

Bhambri, Khang, A., Sita Rani, & Gaurav Gupta, *Cloud and Fog Computing Platforms for Internet of Things*. Chapman & Hall. ISBN: 978-1-032-101507, 2022, https://doi.org/10.1201/9781032101507

Bucea-Manea-Țoniș, R., Martins, O., Bucea-Manea-Țoniș, R., Gheorghiță, C., Kuleto, V., Ilić, M. P., & Simion, V. E. (2021). Blockchain technology enhances sustainable higher education. *Sustainability* 13(22), 12347.

Capece, Guendalina, Nathan Levialdi Ghiron, & Francesco Pasquale (2020). Blockchain technology: Redefining trust for digital certificates. *Sustainability* 12(21 (2020): 8952.

Chen, G. et al. (2018*).* Exploring blockchain technology and its potential applications for education. *Smart Learning Environments* 5(1), 1.

Cheng, J.C., Lee, N.Y., Chi, C., & Chen, Y.H. (2018) Blockchain and smart contract for digital certificate. Presented at the *4th IEEE International Conference on Applied System Innovation, ICASI, Chiba, Japan, 13–17 April 2018*.

Duan, B., Zhong, Y., & Liu, D. (2017) Education application of blockchain technology: learning outcome and meta-diploma. Presented at the *23rd IEEE International Conference on Parallel and Distributed Systems, ICPADS, Shenzhen, China, 15–17 December 2017*.

Emilie, H. Investigating the potential of blockchains. http://blockchain.Open.ac.uk2016-1-22 (2016).

Global, S. G. S. (2016). Sony Global Education Develops Technology Using Blockchain for Open Sharing of Academic Proficiency and Progress Records. Online verfügbar unter. www.sony.net/SonyInfo/News/Press/201602/16-0222E/ (abgerufen am 01.03. 2021).

Gräther, W., Kolvenbach, S., Ruland, R., Schütte, J., Torres, C., & Wendland, F. (2018). Blockchain for education: lifelong learning passport. In *Proceedings of 1st ERCIM Blockchain Workshop 2018. European Society for Socially Embedded Technologies (EUSSET)*.

Gresch, J., Rodrigues, B., Scheid, E., Kanhere, S.S., & Stiller, B. (2019). The proposal of a blockchain-based architecture for transparent certificate handling. In *Proceedings of the 21st International Conference on Business Information Systems, BIS 2018*; Abramowicz, W. & Paschke, A., Eds.; Springer Nature: Basingstoke, UK, 2019; pp. 185–196.

Kamišalíc, A., Turkanovíc, M., Mrdovíc, S., & Heričko, M., A preliminary review of blockchain-based solutions in higher education (2019) In: *Proceedings of the 8th International Workshop on Learning Technology for Education Challenges*, LTEC 2019; Liberona, D., Uden, L., Sanchez, G., & Rodriguez-Gonzalez, S., Eds.; Springer Nature: Basingstoke, UK, 2019; pp. 114–124

Khang, A., et al., IoT equipped intelligent distributed framework for smart healthcare systems, *Networking and Internet Architecture*, 2021, https://arxiv.org/abs/2110.04997v2, doi: 10.48550/arXiv.2110.04997

Khang, A., Geeta Rana, Ravindra Sharma, Alok Kumar Goel, & Ashok Kumar Dubey, The role of artificial intelligence in blockchain applications, *Reinventing Manufacturing and Business Processes through Artificial Intelligence*, 2021, doi: 10.1201/9781003145011

Khang, A., Sita Rani, & Arun Kumar Sivaraman, (2022). *AI-Centric Smart City Ecosystem: Technologies, Design and Implementation*, HB.ISBN: 978-1-032-17079-4 ** EB.ISBN: 978-1-003-25254-2, doi: 10.1201/9781003252542

Kirilova, D., Maslov, N., & Astakhova, T. (2018). Prospects for the introduction of blockchain technology into a modern system of education. *International Journal of Open Information Technologies*, 6(8), 31–7.

Kosba, A., Miller, A., Shi, E., Wen, Z., & Papamanthou, C. (2016). Hawk: The blockchain model of cryptography and privacy-preserving smart contracts. *Security & Privacy*, 839858.

Mikroyannidis, A., Domingue, J., Bachler, M., & Quick, K. (2018). Smart blockchain badges for data science education. In *2018 IEEE Frontiers in Education Conference (FIE)* (pp. 1–5). IEEE.

Nikolskaia, K., Snegireva, D., & Minbaleev, A. (2019) Development of the application for diploma authenticity using the blockchain technology. Presented at the *2019 IEEE International Conference "Quality Management, Transport and Information Security, Information Technologies," IT and QM and IS, Sochi, Russia, 23–27 September 2019.*

Ocheja, P et al. (2018). Connecting decentralized learning records: a blockchain based learning analytics platform. In *Proceedings of the 8th International Conference on Learning Analytics and Knowledge* (pp. 265–269). ACM.

Redman, J. (2018). MIT Media Lab uses the bitcoin blockchain for digital certificates, 12.

Rivera-Vargas, P., & Soriano, C. L. (2019). Blockchain in the university: a digital technology to design, implement, and manage global learning itineraries. *Digital Education Review*, 35, 130–50.

San, A.M., Chotikakamthorn, N., & Sathitwiriyawong, C. (2019) Blockchain-based learning credential verification system with recipient privacy control. Presented at the *2019 IEEE International Conference on Engineering, Technology and Education, TALE, Yogyakarta, Indonesia, 10–13 December 2019.*

Serranito, D., Vasconcelos, A., Guerreiro, S., & Correia, M. (2020) Blockchain ecosystem for verifiable qualifications. Presented at the *2nd Conference on Blockchain Research and Applications for Innovative Networks and Services, BRAINS, Paris, France 28–30 September 2020.*

Sharples, M., & Domingue, J. (2016). The blockchain and kudos: A distributed system for educational record, reputation and reward. In *European Conference on Technology Enhanced Learning* (pp. 490–6). Springer, Cham.

Sun, H. et al. (2018). Application of blockchain technology in online education. *International Journal of Emerging Technologies in Learning*, 13(10), 252–59, DOI:10.3991/ijet.v13i10.9455

Vidal, F., Gouveia, F., & Soares, C. (2019) Analysis of blockchain technology for higher education. Presented at the *2019 International Conference on Cyber-Enabled Distributed Computing and Knowledge Discovery, CyberC, Guilin, China, 17–19 October 2019.*

Williams, P. (2019). Does competency-based education with blockchain signal a new mission for universities? *Journal of Higher Education Policy and Management*, 41(1), 104–17.

Yang, X. et al. (2017). The application model and challenges of blockchain technology in education. *Modern Distance Education Research*, 2, 34–45.

8 Robot Process Automation in Blockchain

R. K. Tailor, Ranu Pareek, and Alex Khang

CONTENTS

- 8.1 Introduction 113
- 8.2 Workflow of Blockchain 115
- 8.3 Components of Blockchain 115
- 8.4 Benefits of Using Blockchain 116
- 8.5 Challenges of Blockchain 117
- 8.6 Advancements of Blockchain 118
- 8.7 Why Need Blockchain Technology 118
- 8.8 Robotic Process Automation and Blockchain 119
- 8.9 Similarities of RPA and Blockchain Technology 120
- 8.10 Advantages of Using RPA and Blockchain Together 121
- 8.11 Opportunities and Challenges of Using RPA and Blockchain Together 121
- 8.12 Uses of RPA in Blockchain 122
  - 8.12.1 Educational Sector 122
  - 8.12.2 Government Sector 123
  - 8.12.3 Online Platforms 123
  - 8.12.4 Business/Corporate Sector 123
- 8.13 Role of RPA in Blockchain Management 123
- 8.14 Conclusion 124
- References 124

8.1 INTRODUCTION

Blockchain is an unchangeable numerical database of commercial operations that can be configured to record anything of value in addition to financial transactions. What this truly implies is keeping track of who owns what, either physical assets (cash, land, and house) or intangible assets like intellectual property, copyright, patent, branding, etc.

Most people associate blockchain with bitcoin, one of the numerous cryptocurrencies. Corporations, governments, and banks have employed records to histories, check account trades when any sales transactions have been documented.

Normally a core authority – the bank, creditors – maintain modifications to the transaction record. Traditionally, document alteration entails the creator issuing the file and then sending it to another for alteration and then forwarding it to a further or back to the author for authentication and managing. It is a framework that secures operational histories, as well recognized as the public building block, in various records it is also known as the "chain" in a network via corresponding nodes.

This storage capacity is also described as an electronic record book. Each operation in that ledger is approved by the owner's electronic sign, which validates the operation and avoids it from being manipulated. It provides a result, whose evidence included in the digital ledger is so reliable (Pankaj Bhambri et al., Cloud & IoT, 2022).

Blockchain is a digital record that has recently gained a lot of interest and popularity technologically, but why has it gotten so popular well leading into it to understand the entire notion. Record keeping of data and transactions is an important part of the business.

Often, this info is handled in-house or passed through a third party such as lawyers, bankers, brokers adding cost, time, or both to the industry or corporate. Fortuitously, blockchain eliminates this time-consuming procedure and enables speedier transaction processing, storing both wealth and time.

Most persons believe that bitcoin and blockchain are interchangeable terms, however, this is not the case. This technology can support an extensive variety of functions in businesses such as supply chain, financing, industrial production, and so on, whereas bitcoin is money that depends on blockchain technology to be protected.

A blockchain is a digital database that saves data electronically (see Figure 8.1). The most well-known aspect of blockchain is its critical function.

Blockchain is quite useful for transaction management. It will take the place of the present transaction management system. There must be an issue if technology replaces the present system. We intend to use blockchain technology to improve

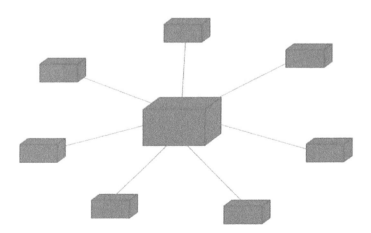

FIGURE 8.1   Blockchain digital database.

Source: Chart is designed by R K Tailor, Ranu Pareek, and Alex Khang.

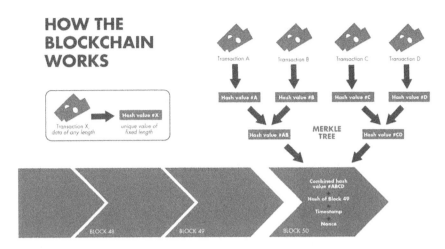

FIGURE 8.2 Workflow blockchain.

Source: www.wasistmalware.de/wpcontent/uploads/2019/11/blockchainworkflow.png/ 17.10pm/07january2022.

banking and bidding. Blockchain is the foundation of bitcoin. It is a public distributed database that houses the encrypted ledger.

8.2 WORKFLOW OF BLOCKCHAIN

Blockchain operates on an associate architecture, which provides sharing and permitting among its authorized members in a closed system. Each member, known as a node, has access to the shared data. This information is synced and modernized to all networks of blockchain members every time a transaction is made to the blockchain.

In today's commercial world, regulations like assignment of the purchase order, offer, and acceptance or other contractual legal agreements that bind other person or party to perform specific acts and remedies control the rules regulating a transaction. Transparency and trust are created when everyone on the blockchain can see and confirm transactions.

The blockchain is built on trust, therefore there is no need for an agent to be engaged in transactions as Figure 8.2.

8.3 COMPONENTS OF BLOCKCHAIN

The blockchain works on various nodes which generates an optimal solution for complex problem with auto triggered robotic devices.

The main components of blockchain are as follows:

a. **Numerical form** – The blockchain is decentralized, and all users and participants in a peer-to-peer network have access to a copy of the whole record. This reduces the need for centralized authority like banks.

b. **Network's numerous participants to attain agreement** – A blockchain relies on the network's numerous participants to attain agreement. Each new block is authenticated and verified by the participants using their computers.
c. **Employs encryption and digital signatures** -Transactions may be tracked back to cryptographic characters, which are supposedly unknown but, with little reverse engineering, can be linked to real-life identities.
d. **Make modifying historical records** – Data that existed previously in a prior block or chain cannot changed in principle unless the protocol's rules allow for such modifications, such as needing the mainstream of its contributors to the consent of this change.
e. **Data-time** – It has a data time for assessment and evaluation with robotic process automation, because blockchain operations are data-time, they may be used to track and verify data.
f. **Robotic Program** – Instructions contained inside blocks, such as "if" this "then" do that "otherwise" do this, permit operations or other operations to be performed just if specific situations are satisfied and can be implemented by adding electronic data. It works on robotic programmed designed for solution and implementation of data driven solution for complex problem.

8.4 BENEFITS OF USING BLOCKCHAIN

It has been discussed above that the blockchain technology is a tool for modern decision making in all business or corporate situations.

It has the following benefits:

a. Transparency and Trust – For many businesses, blockchain provides greater transparency than traditional solutions. Changes on the blockchain are accessible to everyone and not be changed or removed. There is also improved confidence between parties, particularly with nations.
b. Safety and security – Data input cannot be modified, substantially reducing dishonesty. Contracts establish a transparent route, allowing for easy analysis and auditing of each transaction.
c. Cut down costs – Trades might be resolved on a single shared ledger, lowering the expenses associated with authenticating, approving, and checking every operation across different businesses.
d. Enhanced operational speed – The elimination of middlemen and the reimbursement of transactions across numerous centralized third-party approaches provides for faster transaction speeds as evaluated by conventional procedures and procedures.
e. A wide range of applications – On the blockchain, almost everything of value may be recorded.
f. Many firms and sectors are already investigating and creating blockchain-based technologies.

8.5 CHALLENGES OF BLOCKCHAIN

The adoption of technology for institutional and business development matters with their surroundings. It can increase or decrease as per the requirement of their operator.
In adoption of this, these are major challenges:

a. **Non-fiat currency** – The most difficult problem for non-fiat money is regulation. The rate of technological revolution is outpacing the rate of regulatory catch-up. The order of currency evolution has changed from fiat currency to e-money to virtual currency to cryptocurrency. (2)
b. **Governance perspective** – There have already been instances of bitcoin being used for illicit operations such as drug trafficking and money laundering.
c. **Power and mining** – The magic of mining emphasizes a significant difficulty linked with mining power consumption and presents instances of how rising electricity is being invested in mining operations to earn bitcoins.
d. **Scalability** – The scalability challenge of blockchain technology is related to the growth of the people or approval less blockchain. It is optimization and scalability is hotly debated topic.
e. **Cost Reductions** – Robotic process automation can result in cost savings of 50–70 percent.19 Previously robotic automation, one Business Procedure Outsourcing (BPO) facility provider that manage the application for administering insurance profits used a whole-time human worker who could complete the process in an average of 12 minutes. Robotic Process automation provides 24*7 performance at a portion of the cost of human counterparts. Process Automation software finished the procedure in the 1/3rd period and trebled the operation capacity for 1/10th the cost of Full-Time Employees (FTE).
f. **Efficiency & Reduced Downtime** – AI improves workflow and service delivery by improving output and accuracy, decreasing mistakes and phase periods, and eliminating the requirement for repeated instruction. Robots are not like humans, they can easily work twenty-four hours a day, seven days a week. Normally, one robot can perform the job of many full-time employees (FTEs).
g. **Future predication** – Process automation facilitates the collection and business of data, allowing a corporation to forecast future results and enhance existing operations. The analysis indicates areas for improvement, and the enhanced procedures create more detailed data, allowing for further operational improvement and increased efficiency. Future analysis is crucial for achieving monitoring submission, cost-effective growth, and streamlined functions. The analytics may assist in credit risk management by identifying future slow pay and bad debts, as well as giving management information into possible industry economic trends and consideration for policy changes.
h. **Enhanced Performance and Quality** – Even while performing slightly repetitious labor, a human is likely to make 10 mistakes out of every 100 steps. 21

Robots are dependable, steady, and untiring. Bots can complete the same work, similarly, all time, without mistake or deception. AI improves competencies, which increases organizational capacity. One corporation was able to boost managerial efficiency and capability without additional hiring or coaching after adopting automation technologies to assist multiple activities (Khang et al., AI & Blockchain, 2021).

8.6 ADVANCEMENTS OF BLOCKCHAIN

Any software or process works on some technology. Similarly, blockchain is also work on some associate technologies which provides a unique linkage among decision making, peer-to peer analysis and operational performance of each activity and although it facilitates a unique process of perform a task oriented for any specific business or any operational problem.

The main technologies with blockchain and robotic process automation are as follows:

a. **Cryptographical notes** – Private keys and public keys are the two types of cryptography keys. These keys aid in the completion of successful transactions between two parties. The most significant component of blockchain technology is that every individual has two main keys, which are used to create a reliable character. This identification is known as an electronic signature in the bitcoin market, and it is used to sanction and manage operations.

b. **Peer-To-Peer Network Containing a Shared** – Blockchain is a peer-to-peer network in which each peer is autonomous, holds all the data, and any modifications are distributed. When more than one client updates the blockchain, there is a conflict resolution problem in the peer-to-peer network. One strategy is to see who can make the longest chain.

c. **Method of processing, storing operations, and record of the network** – A method of computation used to store network transactions and records. Cryptography keys are made up of two parts: a private key and a public key. These keys aid in the effective completion of transactions between two parties. Each person has these two keys, which they use to generate a secure digital identity reference.

8.7 WHY NEED BLOCKCHAIN TECHNOLOGY

There is always a question on security and safety of data in this digital world. The reason for adoption of blockchain technology is for:

a. **High security** – It makes it harder for other users to alter or change an individual's data without a specific digital signature by utilizing a digital characteristic to complete fraud-free transactions.

b. **Reorganized method** – Traditionally, transactions require the permission of a regulating body like the government or a bank; but, with blockchain, user consensus results in easier, better, and more rapid operations.

c. **The capability of automation** – It is case-sensitive and programmable it may automatically produce logical movements, systematic payments, and events when the trigger requirements are satisfied.
d. **Blockchain has a distributed ledger** – A blockchain is designed to be resistant to data modification. It is defined as "an open, distributed ledger capable of recording transactions between two parties in an efficient, verifiable, and permanent manner."
e. **The ledger is public for all to access** – Blockchains are public ledgers, which means that anybody can view all transactions performed in the past. A human being may use Block Explorers to simply traverse and search your blockchain transaction history. The following are the major components of a transaction: the Transaction ID, the sending and receiving addresses, the related fees, and the transaction's status.
f. **Double spending** – Spending the same amount of money twice is referred to as double-spending. Any transaction, as we all know, can only be handled in two ways. One is offline, and the other is online.
g. **High transaction fees** – The greater the size of a blockchain block, the higher the transaction fees and block rewards. Larger-size blocks take longer to get an agreement on. This is due to how much more difficult bigger blocks maybe for a miner to mine.

8.8 ROBOTIC PROCESS AUTOMATION AND BLOCKCHAIN

Robotic process automation is one of the most recent technologies that is reshaping the industry. Blockchain and artificial intelligence are two topics that have recently made headlines. Because our civilization is an ecosystem, all these technologies are interconnected. Each of these technologies is disruptive, yet corporations are fast to embrace innovative technologies to capitalize on the benefit and saturate deeply into the marketplace.

They are also more likely to integrate technology to get the greatest results for their firm, for example, merging robotic process automation. People frequently ask where robotic process automation fits in with other modern technology. Robotic process automation and blockchain are two developing tools that can revolutionize the way industries operate. These technologies have the potential to have a significant impact. We can see application cases of robotic process automation with blockchain in every industry, including retail insurance, banking, government health care, and so on (Khang et al., IoT & Healthcare, 2021).

Automation is a significant deal nowadays, and businesses rely on it so that they may utilize their spare period to tackle innovative challenges. Creativity and problem-solving are two of the most valuable currencies in today's world. This means that businesses can employ automation to address minor needs, while humans can manage more critical activities that involve understanding, learning, imagination, and strategy.

Using robotic process automation and blockchain, the infrastructure constructed can manage itself. Process automation, for example, is a vital component of every industry. Robotic process automation is an essential component of every industry.

Robotic process automation can manage process automation and thereby reduce expenses on high-volume, repetitive processes that require rules to function.

The automation sector, for example, stands to profit from robotic process automation. Front-office operations may also be automated utilizing robotic process automation. Blockchain, on the other hand, offers a plethora of applications. The greatest impact may be observed in the trade finance industries, which are also making modifications to their procedures and addressing pain areas that were previously disregarded. For example, it is altering the blockchain infrastructure to enable speedier payments and simpler account administration.

8.9 SIMILARITIES OF RPA AND BLOCKCHAIN TECHNOLOGY

Any technological advancement cannot complete without its implication and application. In robotic process automation and blockchain, same technologies are applied to minimize its applicability in various business problems and stages.

The main similarities of robotic process automation and blockchain technology are:

a. **Both are emerging technology** – Blockchain is one of today's innovative technologies, and much revolution and study have just recently begun in this distributed technology and robotic process automation (RPA) is an emerging technology that frequently employs components of artificial intelligence to automate certain processes, intending to free up workers' time to focus on more critical responsibilities.
b. **Depends upon automation** – Robotic process automation (RPA) technology depends or is based upon automation; RPA is a software technology that anybody may utilize to automate digital operations. Although blockchain is thought to be a more secure kind of technology, automation still plays an important part. In one sense, blockchain – regardless of its merits – must be considered in the same light as many other technologies implemented in an IT landscape: it is a system of record that will require installation, operational support, and correct data.
c. **Uses in similar industries** – Blockchain technology and robotic process automation are gaining interest in research due to their effective applications in business, marketing, manufacturing, and finance. However, its use in educational settings is still in its early stages.
d. **Offers security and trust** – Robotic process automation (RPA) can provide effective security defense against unwanted viruses or assaults using AI-powered cognitive RPA bots, which provide several benefits for error reduction and staff efficiency. Cryptography is used to safeguard the records on a blockchain. Network members each have their private key, which is associated with the transactions they do and serves as a personal digital signature. If a record is changed, the signature becomes invalid, and the peer network is immediately notified that something has gone wrong.
e. **Blockchain can act as a network for RPA** – Blockchain, distributed ledgers may offer an immutable record of transactions conducted by RPA-based functions in intelligent process automation.

8.10 ADVANTAGES OF USING RPA AND BLOCKCHAIN TOGETHER

When it comes to information sharing across various IT infrastructures, robotic process automation can allow blockchain. When it comes to monitoring and executing transactions throughout the system, RPA and blockchain can also operate well together.

With the confidence of blockchain, RPA can work easily and offer information inside the environment without worry. With the use of blockchain, transactions may be confirmed and then sent. Furthermore, all data may be saved in the decentralized ledger, guaranteeing immutability to the data recorded. This implies that audit trials will be simple to run for businesses. They may automate the workflow and assure optimal efficiency to suit the needs of their clients.

a. **Governing and agreement managing** – Agreement is a major worry meant for many firms. They must guarantee that their procedure is following the regulatory criteria established by the government under which they are presently working. Things become more difficult if they operate on a global scale. Robotic process automation and blockchain can collaborate to automate repetitive compliance activities. Blockchain plays a significant role in building regulatory and compliance management by providing immutability to the recorded occurrences. All events may also be accessible over the network, allowing for external audits. Blockchain plays a vital role in building regulatory and compliance management by providing immutability to the recorded occurrences. All events may also be accessible over the network, allowing for external audits. This regulatory such as the General Data Protection Regulation (GDPR).
b. **Automated trusted data processing** – Another apparent application is when blockchain works as a core network to manage data transferred by another operation. The decentralized ledger works wonders in this case, as it supports business transactions for automated choices for automated trustworthy data processing in Know your customer (KYC) processes, insurance claim payouts, Human resources (HR) recruitments, and so on.
c. **Customer satisfaction** – Blockchain and robotic process automation (RPA) work together to produce excellent quality client fulfilment with rapid, automated operations that are dependable and trustworthy.
d. **Sustainable Co-worker** – Blockchain and robotic process automation (RPA) are co-workers. Both have the potential to make human companions more sustainable co-workers.

8.11 OPPORTUNITIES AND CHALLENGES OF USING RPA AND BLOCKCHAIN TOGETHER

Robotic process automation (RPA) and blockchain are innovative tools that are seeing quick development on their own. Rapid growth brings its own set of challenges. If you look at blockchain as a standalone technology, it has issues with standards, scalability, and interoperability.

Decentralization is the fundamental issue here! It is difficult to standardize. Scalability is also a concern. However, while next generation blockchain systems are more scalable, they fall well short of the scalability provided by more traditional solutions, such as VISA. Blockchain is now being actively upgraded to address these fundamental issues.

Hyperledger, for example, is attempting to provide a single platform that can be used by many organizations. This will boost both adoption and standardization. Even though there aren't many obstacles to RPA, it nevertheless has a low adoption rate.

Companies are not employing RPA effectively to automate business activities. Most of them are just automating human labor with RPA. They should, however, think strategically about how to get most of the RPA.

In other words, when they implement RPA into their business process, they should increase their value offer. Organizations should also exercise caution when integrating RPA with blockchain technologies.

Furthermore, enterprises should continually conduct a pilot position to assess the effect of RPA and blockchain applications. As a company, you should make certain that you determine important achievement indicators, link with corporate goals, and comprehend the needs.

8.12 USES OF RPA IN BLOCKCHAIN

The use of robotic process automation in blockchain is as flows:

8.12.1 Educational Sector

Blockchain technology and robotic process automation (RPA) may be used in individual educational institutions, groupings of educational institutions, and national and international educational authorities. Anybody who wants to securely retain badges, credits, and certifications – as well as make important educational data available to others. These technologies also make systematic information gathering and arrange systematically.

   a. **International evaluation** – The existing certification system is inadequate for its intended aim. A paper method is vulnerable to lose and even fraud. With an increasingly mobile student and worker populace, a centralized database of qualifications and successes makes sense, whether you're transferring to another academic institution, a new job.
   b. **Blockchain and MOOCs** – The certification problem, on the other hand, is a little hazy. Certificates are issued by each MOOC provider. With a little innovation, safe documentation in the form of an agreement among the main MOOC providers might generate actual demand for MOOCs. It may even allow MOOC accreditation for legitimate degrees. MOOCs are all about decentralization and expanding access, thus it seems to reason that organizer will want to reorganize and expand access to their documentation.

c. **Commercial understanding** – Companies provide extensive guidance to their personnel, but retaining accomplishment is difficult. Existing understanding and ability to manage system technologies. What is required is a more open yet secure system that can be used not only inside but also by workers after they leave a company.

8.12.2 Government Sector

A blockchain – cantered digital authority can safeguard data, improve procedures, and minimize fraud, waste, and abuse while improving trust and accountability. Individuals, corporations, and governments exchange resources using a distributed ledger secured by cryptography in a blockchain-based government model. This approach removes a single point of failure and secures critical citizen and government data intrinsically.

8.12.3 Online Platforms

The development of blockchain technology has the potential to drastically alter control over educational procedures and the distribution of funds in the education sector.

Data about students, professors, and the distribution of educational spending are now held by organizations such as schools and universities. With the use of blockchain, or distributed ledger technology, information will become more accessible, and communication between instructors and pupils will become much simpler.

As a result of this technology, students will be able to select a training course provided in the system without the need for the university or any other body to act as a middleman. At the same time, students will have faster access to knowledge, lowering the amount of work they must do.

8.12.4 Business/Corporate Sector

One of the advantages of using blockchain for business intelligence is that it is no longer associated with technology, but rather with business intelligence, and this is for a reason: blockchain has by far the most extensively utilized business intelligence functions.

Blockchain, which is well-known in the bitcoin business, provides customers with data that cannot be modified due to the availability of records inside networking. Blockchain may be used by a variety of enterprises to revolutionize the way they conduct worldwide economic transactions. Such transactions are especially vulnerable to abuse since both parties are unfamiliar with each other.

8.13 ROLE OF RPA IN BLOCKCHAIN MANAGEMENT

The information saved in these ledgers is more authentic and information is reliable. The tasks remain in the foreground, there is a task, and the swarm solves it on its own in a humanly comprehensive way.

a. **Improve productivity** – Robotic oriented blockchain system is helping in achieving improved productivity and working on the top of the blockchain system. Which is an abstraction of the discrete consensus achievement problem. These technologies made work faster and more accurate.
b. **Reduce the time spend** – Robotic process automation and blockchain technology save time and increase productivity. Reduce the training cost and time to train their employees.
c. **Build up security** – Robotic process automation (RPA)and blockchain technology is a system to distribute crucial information among bots in a safe manner, represented a framework to deal with privacy problems concerning using private data by robots during a human–robot interaction. Require a layer of protection and confirm the confidentiality of the info signified on the data, and with this, the robots can improve Machine Learning (ML) locally with the information they acquired and publish them to the network. Protection against manipulation always plays an important role in all processes of blockchain.
d. **Global information** – Global information can be collected through these technologies and big data can be gathered.
e. **Easier maintenance** – Robotic process automation (RPA) and blockchain have easy maintenance, data can be saved and collected through a single click, and no need for time-to-time observation and maintenance.

8.14 CONCLUSION

To get the greatest results, industries frequently integrate various technologies. The market is highly competitive, and people want more from brands than ever before. That is why businesses must adapt beyond their plan. With technologies integrating more fluidly than ever before, it is now easier for a business to embrace them and design a system that works best for them as well as the end-user. Furthermore, the data obtained by firms is large enough for them to experiment with novel combinations.

All of this is achievable because of the combination of robotic process automation and blockchain technology. Robotic process automation is all about taking command of the various aspects of the business. It is a consensus entity capable of interacting with both the system and the end-users. It is designed in such a manner that it can interact with and interpret real-time data.

Here, the entire staff is tasked with identifying procedures that can be automated, as well as teaching co-workers. So, where does blockchain come in among all of these blockchains that can provide a distributed shared ledger to handle all of the data and information, as well as a smart, trustworthy, and easy trade platform for all parties involved (Khang et al., AI-Centric Smart City Ecosystem, 2022).

REFERENCES

Bhambri, Khang, A., Sita Rani, & Gaurav Gupta, *Cloud and Fog Computing Platforms for Internet of Things*. Chapman & Hall. ISBN: 978-1-032-101507, 2022, https://doi.org/10.1201/9781032101507.

Khang, A., Geeta Rana, Ravindra Sharma, Alok Kumar Goel, & Ashok Kumar Dubey, The role of artificial intelligence in blockchain applications, *Reinventing Manufacturing and Business Processes Through Artificial Intelligence*, 19–38, 2021, https://doi.org/10.1201/9781003145011.

Khang, A., Rani, S., & Sivaraman, A. K. (Eds.). *AI-Centric Smart City Ecosystems: Technologies, Design and Implementation* (1st ed.). CRC Press, 2022. https://doi.org/10.1201/9781003252542.

Khang, A., Sita Rani, Meetali Chauhan, & Aman Kataria, IoT equipped intelligent distributed framework for smart healthcare systems, *Networking and Internet Architecture*, 2021, https://arxiv.org/abs/2110.04997v2, https://doi.org/10.48550/arXiv.2110.04997.

9 A Novel Approach to Cryptography

Deep Learning-based Homomorphic Secure Searchable Encryption for Keyword Searches in the Blockchain Healthcare System

T. B. Sivakumar and Hasan Hussain

CONTENTS

9.1	Introduction	127
9.2	Hyperledger Composer	128
9.3	Related Work	130
9.4	System Architecture	130
9.5	System Implementation	131
9.6	Discussion	132
9.7	Conclusion	133
References		134

9.1 INTRODUCTION

Nowadays, information interchange between patients, corporate entities such as multiple hospital systems, pharmaceutical companies, and other entities has long been a part of the healthcare management system. Nonetheless, the patient-driven personal health record (PHR), in which health information sharing is handled by the patient, has received considerable interest.

PHR interoperability, in general, entails new requirements and problems in terms of technology, incentives, security and privacy, and governance, all of which must be addressed in order to resolve data sharing issues. The application of blockchain technology in healthcare management systems (Khang et al., IoT & Healthcare, 2021) can provide a variety of management systems for healthcare, including those that handle PHR, extremely sensitive data, and PHR entities. A growing number of medical data assesses acts such as producing, sharing, and modifying information items, making it harder to track criminal activity and security breaches (Roehrs et al., 2017).

A patient health record (PHR) is a system for digitally preserving a patient's medical information. It must enable for adequate access control in order to monitor, track, and restrict personal health information. The PHR stores detailed health information about a patient, such as appointment dates, prescription drug plans, allergy reports, immunization records, lab findings, and so on. Malicious users can obtain the patient's health information during the traditional emergency access of the PHR practice, while the Emergency Team (EMT) performs activities on the medical records. Most crucially, in the traditional system, an auditing trail or activity tracking system is required, as well as the ability for the patient to designate rights for accessing the PHR (Ahvanooey et al., 2019).

We present a unique management system that utilities the shared and changeless distributed ledger and is based on a blockchain network. Blockchain is a technology that uses shared servers to create a secure, difficult-to-tamper ledger. When a transaction gets endorsed, the transaction is difficult to change lawfully due to the blockchain network-based system's capabilities. To gain consensus on the new event for the blockchain, it employs a number of consensus algorithms. In general, blockchain analyzes the security standards stated previously to assure the dependability of created records, also known as blocks, which include events. It also needs to enhance accountability by empowering authoritative participants' entry and access control (Señor et al., 2012).

The blockchains most important feature is auditing. When a transaction is completed, the current block records it with a timestamp, and the system's participant keeps track of earlier event activities. It keeps track of all transactions. Individuals or medical organizations who require tamper-proof account records will benefit from this method (Tang et al., 2006).

Our solution is based on the Hyperledger composer blockchain, which could provide an efficient mechanism for resolving malicious access to the PHR, i.e., an off-chain extendable and scalable data storage and a person-centered mobile and web edge. The blockchain is used in this paradigm to ensure non-repudiation, accountability, and tamper-proof features (Krukowski et al., 2015).

The delegate re-encryption approach is used to suggest an access control tool for granular access authority. The suggested solution makes use of smart contracts, which allow the PHR's owner to establish rules for an EMT or staff member (certified physician) to get permission to access current information from the PHR while taking into account the time limit (Adida et al., 2006).

In a regular situation, the patient and their family physician can easily access the system via a web browser and mobile interface in a hyperledger composer application.

9.2 HYPERLEDGER COMPOSER

The Linux Foundation supported Hyperledger Fabric initiatives, one of which is the Hyperledger Composer (see Figure 9.1). The business network archive (BNA) is a Hyperledger Composer feature that was inherited from the Hyperledger Fabric blockchain (Zhang et al., 2018).

A Novel Approach to Cryptography

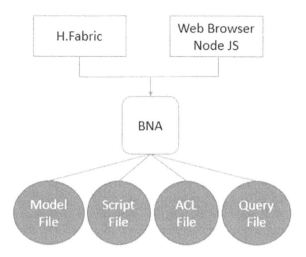

FIGURE 9.1 Hyperledger composer – Architecture.
Source: Architecture is designed by T.B Sivakumar and Hasan Hussain.

The business network is made up of members, who are linked together by their unique identifiers, as well as assets generated by the system; transactions establish asset exchange.

These rules entail executing smart contract transactions, and all of the transactions are eventually saved on the ledger. Participants, assets, and transactions are the three key components of the model file. The participants are the system's end-users, and they can interact with assets and communicate with one another through transactions (Benchoufi et al., 2017).

The variables saved in the network are usually referred to as assets. Transactions are the system's goals, and they're used to keep the setup up to date. Multiple transaction functions in the system are determined by the Script file in the business network. It is made up of Java Script (JS) and deals with business logic, such as which user standards to follow and which assets to share. The access control list (ACL) defines the different ranges of access that each participant has in the network (Wang et al., 2018).

The participants' goal is set in the ACL file, and it determines their performance in creating, reading, updating, or deleting assets. The Query file outlines how the system's queries are put together and used. These are fixed to extrapolate historian transactions based on all prior transaction records in the network (Puthal et al., 2018).

The historian record is a registry list that provides the history of transactions and events done on the system and is provided by the historian record. The historian record is updated when the transaction is being executed, saving a history of all transactions inside a corporate network (Delmolino et al., 2016).

The participants are involved in submitting the transactions with their identities, and historian record assets can be accessed using composer queries to request specific records.

9.3 RELATED WORK

In this section, we review the current state-of-the-art healthcare management systems and discuss their benefits and drawbacks.

Guy Zyskind et al. presented the Enigma privacy platform, which uses blockchain to manage access control and auditing logs, as well as privacy and security goals including a tampering proof transaction record (Zyskind et al., 2015). Enigma uses a multi-party computing approach and a verified secret exchange mechanism to ensure data privacy. Researchers believe that Enigma eliminates the need for a trusted third-party platform, allowing personal data control to be done privately (Christidis et al., 2016).

Hussein proposed a data-sharing method based on blockchain to handle the issues of access control with the blockchain, such as autonomy qualities and immutability (Hussein et al., 2018). The authors used a Discrete Wavelet Transform (DWT) and a genetic algorithm to improve the queuing optimization technique in this work. As a result, it generates a cryptography key for access control and immunity, allowing users to be authenticated quickly.

Dagher et al. proposed a blockchain-based architecture for delivering dynamic, interoperable, and secure access to medical records while safeguarding patients' personal data. Researchers used the Ethereum blockchain to create this system, which included smart contracts for access control and data obfuscation, as well as cryptographic measures for added protection (Dagher et al., 2018).

Chen et al. devised a mechanism for storing blockchain-based personal medical data in the cloud (Khang et al., IoT & Healthcare, 2021). They use blockchain to create a storage supply chain in which all transactions are verified, immutable, and transparent. This system outlined the permissions for three categories of transactions, as well as the block formation and principal function of the medical blockchain. They also built a service framework for sharing medical records that safeguards medical data management apps while without infringing on privacy policies (Chen et al., 2019).

Zhang et al. proposed a blockchain-based secure and privacy-preserving personal health information sharing protocol for e-Health diagnosis improvements. They also explained the blockchain consensus mechanism, which is a proof-of-conformance system for creating validated blocks (Zhang et al., 2018).

Furthermore, the researchers used public-key encryption with a blockchain-based keyword search. After receiving trapdoors from the patient, a doctor is able to search and access the expected history of health records in order to improve the diagnosis. Furthermore, they claimed that this eHealth system achieves security, privacy preservation, and secure medical data searches.

The state-of-the-art studies cited above are based on blockchain-based health record sharing and access control policies. Despite this, they do not use PHR in an emergency. For safeguarding data privacy and auditing trial in emergency access for PHR, we employed a Hyperledger Composer and Fabric (Chen et al., 2019).

9.4 SYSTEM ARCHITECTURE

We present the suggested emergency access control management system in this section, which makes use of blockchain technology to protect PHR data privacy.

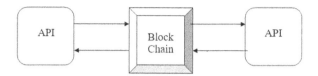

FIGURE 9.2 Proposed framework.

The nodes on the blockchain network share all of the data. Using the Hyperledger Composer-SDK and NodeJS, we create a system that generates a time-stamped record for all transactions on the network without involving a PHR owner or any other third party.

Furthermore, in Figure 9.2, we show how the suggested architecture, which uses the Hyperledger composer blockchain to provide access control scenarios of PHR data in an emergency, works. We start by defining the entities that are involved in the construction process.

Permissions and smart contracts govern all of the activities that effect data extraction from the Ledger as Figure 9.2.

Patient refers to a participant who owns the PHR data. The PHR data access control policies are defined by a patient. A doctor is a participant who has authority to log into the system if the patient has given him such permission. As a family doctor or main physician, the PHR owner must set the policy of access control permission in a smart contract.

A participant who asks emergency access permission when the patient is in an emergency is known as an Emergency Doctor. The suggested framework uses an API to offer access to the emergency doctor based on the patients' rules, whether or not he is allowed to view the PHR data.

Consensus is a mechanism in our system that performs the following basic functions for authorizing transactions and confirming the patient's policies. When the transaction is complete, the Consensus acknowledges the results and upgrades the main shared ledger to ensure that the results are consistent.

A ledger is the result of tampering with secret records for all transactions. Transactions are the results of smart contracts or requests sent in by users. The completion of each transaction is a k-v pair connected to the state of creates, updates, or deletes.

9.5 SYSTEM IMPLEMENTATION

In this section, we use Hyperledger Fabric and Hyperledger Composer to implement the proposed model. During our tests, we assume that the user (client) information is collected from JSON, and that the required information is obtained by using the Rest Client, i.e., Postman server.

Every server was created in an Amazon Web Server (AWS) virtual environment Elastic Compute Cloud (EC2) instance (Pankaj Bhambri et al., Cloud & IoT, 2022), which runs on the same local personal computer with Ubuntu Linux 18.04.1, a single vCPU @ 2.00 GHz, and 32 GB RAM, as shown in Table 9.1.

TABLE 9.1
Implementation Development Environment

Component	Description
CPU	Single vCPU @ 2.00 GHz
Operating System	Ubuntu Linux 18.04.1 LTS
Memory	32 GB
Hyperledger Fabric	Version 1.2
Docker	Version 1.12.1
Oracle Virtual Box	Version 5.1.22
Docker	Compose Version 1.5.

Table 9.1 shows how to create the Business Network Definition, we used the Hyperledger composer playground.

A patient-centric user interface, a permissioned blockchain, and off-chain storage are the three components of our suggested architecture. The Hyperledger Composer was also used to create the Business Network Archive (BNA), which describes the network's features and capabilities. On the Hyperledger Fabric instance, Hyperledger Composer is also utilized to archive the business network.

9.6 DISCUSSION

Our system protects the patient's privacy by allowing granular access control over his or her PHR data to be defined. Furthermore, by merging smart contracts, it addresses access control management. The proposed model performs in the Hyperledger composer network depending on the identities of the given participants. As a result, malevolent people will be unable to access the PHR data.

In the HF, channels are built based on access regulations that control who has access to the channel's storage, which include smart contracts, transactions, and ledger states. As a result, these channels are made up of nodes that specify PHR privacy and confidentiality.

The PHR data is protected by our suggested framework against ransomware and other security breaches such as unauthorized access in Figure 9.3.

The emergency doctor was unable to access the PHR data after the time limit on his/her access data had expired. The use of blockchain technology simplifies and reduces the cost of implementing the system.

By keeping access control lists and logs directly on the blockchain, the system improves security, privacy, and auditability. Before granting access to a user, each attempt to access a record is checked in the access control list and logged. The technology establishes a new standard for handling emergency access control and auditing among several participants.

Experiments show that our approach outperforms the standard emergency access method in terms of efficiency.

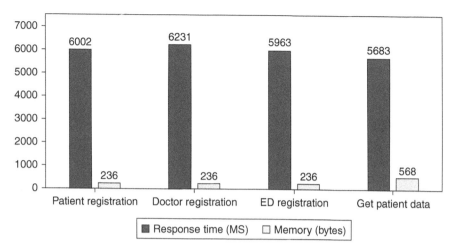

FIGURE 9.3 Performance evaluation.

TABLE 9.2
Analysis of the Proposed Framework vs the State-of-the-Art Systems

Healthcare System Name	Patient Identity	Immutability	Data Auditing	Smart Contracts	Access Control
Our framework	✓	✓	✓	✓	✓
Xiao et al., 2016 [34]	✓	✓	✗	✗	✗
Hussein et al., 2018 [38]	✓	✓	✗	✗	✗
Dagher et al., 2018 [39]	✓	✓	✗	✓	✗
Chen et al., 2019 [40]	✓	✓	✗	✗	✗
Zhang et al., 2018 [41]	✓	✓	✓	✓	✗

In addition, after recovering from the emergency situation, patients receive historical records for the audit trail and check the access control policies to ensure that their PHR data is not compromised.

This chapter describes how a blockchain framework (Khang et al., AI & Blockchain, 2021) can be used to improve the auditing and privacy of PHR systems as shown in Table 9.2.

9.7 CONCLUSION

Not only does our proposed architecture provide security controls for managing access permissions to PHRs during an emergency, but it also allows the health management system to reduce the need for emergency contact. However, there are several shortcomings in Healthcare 2021, 9, 206 15 of 17 that should be addressed in future research.

We should test our framework by engaging diverse groups of participants and taking their feedback into account throughout the maintenance stage, since it is still in the prototype stage. Furthermore, because PHRs are exchanged/shared among multiple participants, a standard such as HL7 FHIR is required to ensure data sharing implementation security (Khang et al., AI-Centric Smart City Ecosystem, 2022).

REFERENCES

Adida, B., & Kohane, I.S. 2006. GenePING: Secure, scalable management of personal genomic data. *BMC Genom.* 2006, 7, 93.

Ahvanooey, M.T., Li, Q., Hou, J., Rajput, A.R., & Yini, C. 2019. Modern text hiding, text steganalysis, and applications: A comparative analysis. *Entropy* 2019, 21, 355.

Benchoufi, M., & Ravaud, P. 2017. Blockchain technology for improving clinical research quality. *Trials* 2017, 18, 335.

Bhambri, Pankaj, Sita Rani, Gaurav Gupta, & Khang, A., 2022. *Cloud and Fog Computing Platforms for Internet of Things*. Chapman & Hall. ISBN: 9781032101507, doi: 10.1201

Chen, Y. et al., 2019. Blockchain-based medical records secure storage and medical service framework. *J. Med. Syst.* 2019, 43, 5.

Christidis, K. & Devetsikiotis, M. 2016. Blockchains and smart contracts for the Internet of Things. *IEEE Access* 4, 2016, 2292–2303.

Dagher, G.G., Mohler, J., Milojkovic, M., & Marella, P.B. 2018. Ancile: Privacy preserving framework for access control and interoperability of electronic health records using blockchain technology. *Sustain. Cities Soc.* 39, 2018, 283–97.

Delmolino, E. et al. 2016. Step by step towards creating a safe smart contract: lessons and insights from a cryptocurrency lab. In *Proceedings of the International Conference on Financial Cryptography and Data Security, Christ Church, Barbados*, February 26, 2016.

Hussein, A.F., Arunkumar, N., Ramirez-Gonzalez, G., & Abdulhay, E., Tavares, J.M.R., & De Albuquerque, V.H.C. 2018. A medical records managing and securing blockchain based system supported by a genetic algorithm and discrete wavelet transform. *Cogn. Syst. Res.* 52, 2018, 1–11.

Khang, A., Geeta Rana, Ravindra Sharma, Alok Kumar Goel, & Ashok Kumar Dubey, 2021. The role of artificial intelligence in blockchain applications, *Reinventing Manufacturing and Business Processes Through Artificial Intelligence*, 19–38, https://doi.org/10.1201/9781003145011

Khang, A., Sita Rani, Meetali Chauhan, & Aman Kataria, 2021, IoT equipped intelligent distributed framework for smart healthcare systems, *Networking and Internet Architecture*, 2021, https://arxiv.org/abs/2110.04997v2, https://doi.org/10.48550/arXiv.2110.04997

Khang, A., Rani, S., & Sivaraman, A.K. (Eds.). 2022. *AI-Centric Smart City Ecosystems: Technologies, Design and Implementation* (1st ed.). CRC Press. https://doi.org/10.1201/9781003252542

Krukowski, A., Barca, C.C., Rodríguez, J.M., & Vogiatzaki, E. 2015. Personal Health Record. Cyberphys. Syst. Epilepsy Brain Disord.

Puthal, D., Malik, N., Mohanty, S.P., Kougianos, E., & Yang, C. 2018. The blockchain as a decentralized security framework [future di-rections].*IEEE Consum. Electron. Mag.* 2018, 7, 18–21.

Roehrs, A., Da Costa, C.A., Da Rosa Righi, R., & De Oliveira, K.S.F. 2017. Personal health records: A systematic literature review. *J. Med. Internet Res.* 2017, 19, e13.

Señor, I.C., Fernández-Alemán, J.L., & Toval, A. 2012. Are personal health records safe? A review of free web-accessible personal health record privacy policies. *J. Med. Internet Res.* 2012, 14, e114.

Tang, P.C., Ash, J.S., Bates, D.W., Overhage, J.M., & Sands, D.Z. 2006. Personal health records: Definitions, benefits, and strategies for overcoming barriers to adoption. *J. Am. Med. Inform. Assoc.* 2006, 13, 121–6.

Wang, S., Zhang, Y., & Zhang, Y. 2018. A blockchain-based framework for data sharing with fine-grained access control in decentralized storage systems. *IEEE Access* 2018, 6, 38437–50.

Zhang, P., White, J., Schmidt, D.C., Lenz, G., & Rosenbloom, S.T. 2018. FHIRChain: Applying blockchain to securely and scalably share clinical data. *Comput. Struct. Biotechnol. J.* 16, 267–78.

Zyskind, G., Nathan, O., & Pentland, A. 2015. Enigma: Decentralized computation platform with guaranteed privacy. arXiv 2015, arXiv:150603471.

Zhang, A. & Lin, X. 2018. Towards secure and privacy-preserving data sharing in ehealth systems via consortium block-chain. *J. Med Syst.* 2018, 42, 140.

10 Design and Implementation of a Smart Healthcare System Using Blockchain Technology with A Dragonfly Optimization-based Blowfish Encryption Algorithm

Shivlal Mewada, Dhruva Sreenivasa Chakravarthi, S. J. Sultanuddin, and Shashi Kant Gupta

CONTENTS

10.1	Introduction	138
10.2	Problem Statement	140
10.3	Proposed Work	141
	10.3.1 Dataset Description	142
	10.3.1.1 Patients Generated Data	142
	10.3.1.2 Health and Clinical Records Data	142
	10.3.2 Data Preprocessing Using Normalization	142
	10.3.3 Data Validation Using Smart Contracts	144
	10.3.4 Dragonfly Optimization Based Blowfish Encryption Algorithm	144
	10.3.4.1 Key-Expansion	145
	10.3.4.2 Data Encryption	145
	10.3.4.3 F Function in Data Encryption	145
	10.3.5 Blockchain Mechanism	146
	10.3.5.1 Verification of Data Using Proof of Stake (PoS) Consensus Protocol	147
	10.3.5.2 Decryption of Data and Access Granting	148
	10.3.5.3 End Users	148

DOI: 10.1201/9781003269281-10

10.4　Performance Analysis 148
　　　10.4.1　Encryption Time 148
　　　10.4.2　Security Level..148
　　　10.4.3　Communication Cost 149
　　　10.4.4　Computation Cost...150
　　　10.4.5　Time Consumption 151
10.5　Conclusion 151
References..151

10.1　INTRODUCTION

Smart healthcare was always a major concern that must be handled in any way possible in light of the world's technological breakthroughs. The only thing that counts is that medical services be improved in terms of organization, credibility, method, and effectiveness, and also patient nourishment and caring.

People are growing increasingly unwilling to seek personal healthcare until a catastrophic catastrophe develops in today's atmosphere. This is often seen as a part of over-engagement with the customary hectic existence and tailored lifestyle pattern.

As a result, if a system is developed that measures or detects typical anomalies in a human's health suite and can submit to the person's assigned personal health care supervisor, the entire situation will be much more advantageous, and an easy discussion on the patient can be done at the right time and within a safe duration.

Technological innovations like the Internet of Things and big data have assisted in the expansion and development of healthcare all through the world, and also the development of a smart healthcare system.

Smart healthcare is a medical system, which builds health and medical operations and optimizes administration employing Internet of Things, data transfer, and interchange techniques, with medical data in the cloud at its core.

Although the smart healthcare industry is progressing, there are still challenges with data and system security. A healthcare system that is smoothly connected with a cloud-based system is referred to as smart healthcare. Decentralization, digitization, accessibility, reachability, and other aspects of a smart healthcare system are all part of the concept. The conventional healthcare system consists of a vast network of people, systems, and technologies (Pankaj Bhambri et al., Cloud & IoT, 2022).

Finances, legalities, logistics, employees, and medical records are just a few of the components. All of the systems and procedures in smart healthcare are made more convenient for all parties involved. Various challenges in the smart healthcare are depicted in Figure 10.1.

Blockchain, the core innovation of the 4th industrial revolution, has qualities such as decentralization, privacy, tamper resistance, and traceability. The integration of blockchain and smart healthcare could address conventional smart healthcare's problem areas in information exchange, information security, and confidentiality management, improve user-centered smart healthcare systems, and generate a multi-party clinical coalition sequence encompassing government, businesses, and persons to enhance smart healthcare's industrial improvement. Blockchain has grabbed the curiosity of the entire sector at this moment.

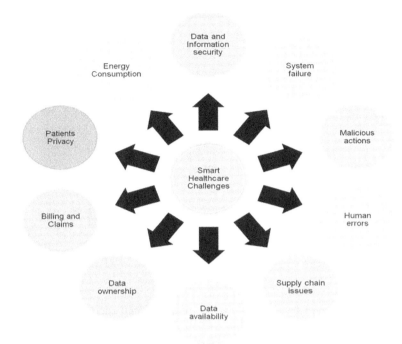

FIGURE 10.1 Various challenges in smart healthcare.

The present study findings include basic blockchain bottom operations, blockchain access control, long-term validation assessment of blockchain signature, and so on. The research into the application of blockchain in smart healthcare, on the other hand, is still in its early phases.

The bulk of current research aims to integrate blockchain with current information technologies to build a new data platform or network, like the creation of a blockchain-based computerized healthcare system or a blockchain-based information privacy protection system.

There are also some studies that look into the contribution of blockchain in the long-term distribution network of smart healthcare, like tracking counterfeit and sub-standard drugs with blockchain, tracking the operational environment of pharmaceutical devices with blockchain, and integrating and upgrading to a supply chain network with blockchain.

Regardless of the fact that the above-mentioned research has done significant research on the application of blockchain in the smart healthcare field, the smart healthcare development system under blockchain remains vague and requires systematic research.

Moreover, the applicability of blockchain in the smart healthcare system is hard to define in detail since it will cooperate with all matters in the smart healthcare supply chain all through the application procedure, and multiple subjects would need to cooperate and affect one another. The basic structure of blockchain based healthcare system is shown in Figure 10.2.

FIGURE 10.2 Blockchain in healthcare.

The healthcare industry is no longer an outcast, unaffected by the wonders of current technology advancements. With technical breakthroughs and the convergence of computer science, electronics, and associated technologies, it is now feasible to build a smart healthcare environment in which all medical entities are linked and can interact with one another.

With the introduction of blockchain, the technology has recently taken over total access, transaction, and storage management. Blockchain has also showed tremendous traction and promise in a variety of industries, including retail, supply chain management, finance, healthcare, and so on.

In healthcare, the safety and confidentiality of information, which is often utilized by a number of stakeholders for a variety of purposes, is always a big issue. As a consequence, this approach is a feasible choice for safeguarding data against misuse while also maintaining a high degree of trust among the many stakeholders.

10.2 PROBLEM STATEMENT

Smart healthcare recognizes the engagement between patients and medical staff, medical institutions, and medical devices by creating a health records regional medical information structure and utilizing the most innovative Internet of Things technology. This allows the medical industry to gradually achieve information.

Medical data sharing is a crucial step in making the medical system smarter and enhancing the quality of medical services. Nonetheless, cross-institutional patient data interchange is still a work in progress, and the blockchain is a great way to address this problem right now.

Nevertheless, there is a scarcity of thorough research on blockchain technology's usage in the intelligent medical industry as a whole… Hence, this chapter proposes a novel model for smart healthcare system using blockchain.

10.3 PROPOSED WORK

This chapter proposes a system to overcome the issues and challenges faced by the classical SHS. This chapter discusses about the blockchain based smart healthcare system.

To enhance the security of the healthcare data during transmission in the blockchain network, Dragonfly Optimization based Blowfish Encryption Algorithm (DO-BEA) is proposed. The schematic representation of the proposed method is shown in Figure 10.3.

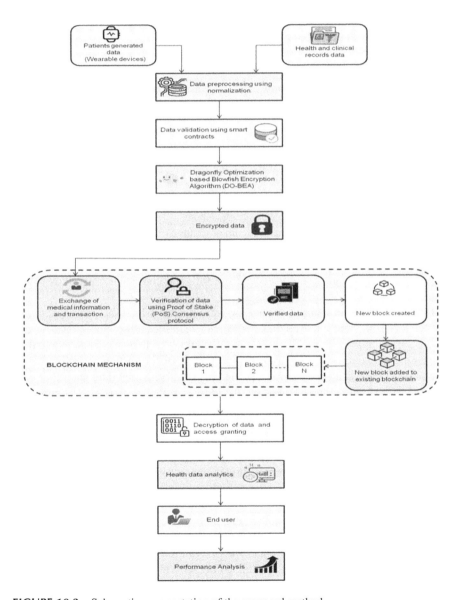

FIGURE 10.3 Schematic representation of the proposed method.

10.3.1 Dataset Description

10.3.1.1 Patients Generated Data

Doctors utilize wearable devices to keep track of their patients' symptoms in real time. These are just wristbands and timepieces that may be worn on the wrist or on the body. These wearable gadgets are made up of tiny sensors that detect the surroundings as well as the patients' vital signs (like blood pressure, heart rate, pulse rate, ECG, humidity, temperature etc).

The physicians and clinicians are alerted in real time when these parameters deviate from the acceptable boundaries. The notice is sent to the relevant doctor in the form of an SMS as well as on smart devices.

10.3.1.2 Health and Clinical Records Data

The patient's continual monitoring creates a vast quantity of data. This data isn't always relevant in real time, but when properly evaluated, it may provide significant insights.

As a result, the cloud stores all of the data collected by the smart wearable gadget and other monitoring devices. Before the data is stored, it must be encrypted using the blockchain technology to protect it from data theft and eavesdropping.

10.3.2 Data Preprocessing Using Normalization

Typically, healthcare databases are made up of a range of heterogeneous data sources, and the data extracted from them is different, partial, and redundant, all of which have a significant impact on the final mining outcome. As a result, healthcare data must be preprocessed to guarantee that it is accurate, full, and consistent, as well as having privacy protection.

Data normalization is a pre-processing technique in which the data is scaled or altered to ensure that each characteristic contributes equally. Normalization is a scaling, mapping, or preprocessing method that allows us to create a new range from an existing one. It may be quite useful for predicting or forecasting purposes.

Normalization is a process that rescales or transforms raw data such that each feature has a consistent contribution. It addresses the existence of dominating features and outliers, two major data concerns that obstruct the learning process of machine learning algorithms. Many approaches for normalizing data within a specific range based on statistical measures from raw (un-normalized) data have been developed.

The Min-Max and Z-score normalization approaches are employed in this study. These approaches are classified depending on how specific raw data statistical features are utilized to normalize the data.

Min-Mix Normalization is a method that delivers linear transformation on an initial range of information. Min-Mix Normalization is a method that maintains the connection between the original data. Min-Max normalization is a basic approach that can precisely fit data inside a predefined border with a pre-defined boundary.

As per Min-Max normalization technique,

$$L' = \left(\frac{L - \text{minvalue of } L}{\text{maxvalue of } L - \text{minvalue of } L} \right) * (O - G) + G \quad (10.1)$$

where L contains Min-Max Normalized data one
If pre-defined boundary is [G, O]
If L is the range of original data & M is the mapped one data.

The parameter Z-score Normalization is a method that generates normalized values or ranges of data from unstructured information utilizing notions such as mean and standard deviation. As a result, the unstructured data may be normalized using the z-score parameter, as seen in eqn. (10.2):

$$e'_j = \frac{e_j - \bar{X}}{std(X)} \quad (10.2)$$

where
e'_j is Z-score normalized one values.
e_j is value of the row X of jth column

$$std(X) = \sqrt{\frac{1}{(k-1)} \sum_{j=1}^{k} (e_j - \bar{X})^2} \quad (10.3)$$

$$\bar{X} = \frac{1}{k} \sum_{j=1}^{k} e_j \text{ or mean value} \quad (10.4)$$

Assume we have five rows, A, B, C, D, and E, each with separate variables or columns that begin with the letter "j." As a result, the z-score approach may be used to determine the normalized values in each row above. If all of the values in a row are identical, the standard deviation of that row is zero, and all of the values in that row are set to zero. The z-score, like the Min-Max normalization, indicates the range of values between 0 and 1.

Decimal Scaling is the approach that offers the range between -1 and 1. As a result of the decimal scaling method,

$$e_j = \frac{e}{10^p} \quad (10.5)$$

where
e^j is the scaled values
e is the range of values
p is the smallest integer $\text{Max}(|e^j|) < 1$

10.3.3 DATA VALIDATION USING SMART CONTRACTS

The data and information collected by various sensors might take many different forms. In order to conduct effective analysis, this data must be converted into a single standard format.

Smart contracts are being created and digitally signed by all parties in the system. Blockchain is considered tamper-proof because to its immutable structure, offering complete security, consistency, and transparency, as well as user privacy. Ethereum is used to represent the storage of medical records by storing each node in a network. Developing an intelligent healthcare system requires rigorous record-keeping (Khang et al., IoT & Healthcare, 2021).

In this case, all records, information, ownership, and permissions are encrypted and processed using signed instructions. This overcomes the issue of protecting sensitive patient information. Each party must provide their consent before any information may be shared. Because smart contracts feature inherent data authorization limits and unique IDs, they are excellent for storing medical data.

10.3.4 DRAGONFLY OPTIMIZATION BASED BLOWFISH ENCRYPTION ALGORITHM

Dragonfly optimization based Blowfish Encryption Algorithm (DO-BEA) is a novel approach used in this work to generate the revolutionary key. The main goal of this algorithm is to improve the security and privacy of healthcare data including data and images.

In this chapter, MATLAB is used to create a single objective function for this optimization. It offers the work's result as the finest solution in the world. Files may also be encrypted using the supplied encryption technique. The file type includes executable files, portable format files, text files, and word documents, as well as other files with increased security.

The method for encrypting files is the same as for encrypting text data. Regardless of their original format, the files' contents are converted to binary blocks and dragged into a text page. This is a step in the file encryption process that is optional. The text data encryption procedure will begin immediately once the data has been translated and transferred to a text document.

During the decryption method, the ciphered data is reinserted in the text file, and the decryption takes place. In the event of link files, library files, or image data, the output is delivered as damaged information for encryption.

To make the algorithm stronger and tougher, many keys are used in the encryption process. The encryption code is automatically reinforced when the number of keys is raised. Furthermore, the key is generated from the supplied data, making retrieval difficult.

Furthermore, a sub-optimal key is developed and used for encryption, making the procedure so onerous that an intruder would be unable to access the data. The file will be decrypted by the authorized recipient who has access to the key.

Blowfish is a symmetric encryption technique, which means it encrypts and decrypts communications using the same secret key. Blowfish has a 64-bit block

Design and Implementation of a Smart Healthcare System

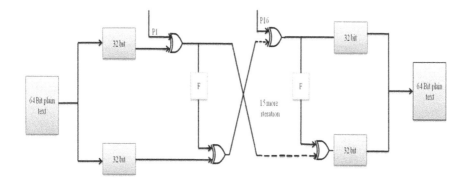

FIGURE 10.4 Blowfish encryption.

length, therefore messages that aren't multiples of eight bytes should be sheltered. Blowfish is split into two categories:

- Key-expansion
- Data encryption

10.3.4.1 Key-Expansion

The entered key is converted into numerous sub key arrays totaling 4168 bytes during the key expansion process. The P array is made up of 18 32-bit boxes, while the S-boxes are made up of 4 32-bit arrays with 256 items each.

The initial 32 bits of the key are XORed with P1 after string setup (the first 32-bit box in the P-array). The second 32 bits of the key are XORed with P2, and so on until all 448 bits are used.

10.3.4.2 Data Encryption

The information is used utilizing 64-bit plain text and then encoded to 64-bit cipher text at this point. The 64 bits of data files are divided into 2 32-bit as left parts and right halves, as shown in Figure 10.4.

The result of XORing each 32-bit with P-cluster is sent to the function (F). Then, with ease, finish the XOR task for identically left bits and the corresponding 32 bit right bits. This technique continues until the 16th round is completed.

10.3.4.3 F Function in Data Encryption

The F function generates four 32-bit S-boxes, each with 256 elements. The fundamental 32 bit left halves are partitioned into four 8 bit blocks, like v, u, c, and k, in the unique blow fish approach. Equation depicts the formulation for the F_t function (10.6).

$$U(RK) = \left(\left(D_{a1} + D_{a2}, \mod 2^{32}\right) \oplus D_{a3,c}\right) + D_{a4,h} \mod 2^{32} \qquad (10.6)$$

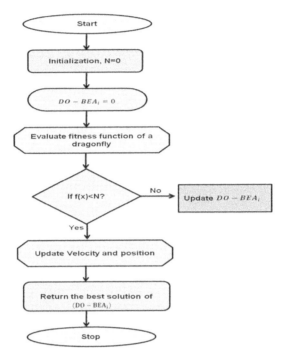

FIGURE 10.5 Flowchart of DO-BEA.

Data is converted into binary blocks, which reduces processing time. The optimization strategy will boost execution speed while also reducing storage capacity. The space complexity is a serious problem to address since the method is built for the application service to be mounted to the cloud.

As a result, transforming data to binary blocks saves space, and computations on this blocks structure will result in a first-class destruction if any errors are discovered. The suggested model's block size ranges from 8 X 8 bits to 256 X 256 bits, while the key size ranges from 64 to 256 bits. The flowchart for Dragonfly Optimization based Blowfish encryption Algorithm is illustrated in Figure 10.5.

10.3.5 BLOCKCHAIN MECHANISM

The encoded transaction data is saved in the form of unchangeable storage blocks. Only authorized personnel have access to the information. We can use blockchain technology to develop privacy-preserving and inherently secure data interchange platform that enables for quick access to archived and real-time patient data to collaborating institutions through smart contracts, removing the need for data reconciliation entirely.

A blockchain is a decentralized control and access system in which each participant has a vested interest but no single administrator. Everyone has the same power and rights.

10.3.5.1 Verification of Data Using Proof of Stake (PoS) Consensus Protocol

PoS protocol has gotten a lot of interest since PoW protocol requires a lot of resources and computing resources tend to be centralized. PoS protocol believes that wealthier equity owners are more eager to maintain the system's consistency and security.

In specifically, when the PoS system has validated the packing requirement, the node may be chosen as a representative to propose a new block at the start of each round. After getting the longest valid blockchain, the representative proposes a new pending block and broadcasts the new blockchain he constructed while waiting for confirmation.

The PoW mechanism (Khang et al., AI & Blockchain, 2021) reselects the representative at the start of the following round to validate the previous round's results. Since the longest blockchain is still legitimate, honest representatives will continue to operate behind it.

Miners in a PoS system get packing opportunity once their pseudorandom hash function meets the following discrepancy, analogous to the PoW algorithm (10.7).

The distinction from the PoW is that even if challenge b could be performed is solely determined by the node's equity and has nothing to do with the node's computational capacity.

The more a node's ownership, the more likely it is to be chosen as an indicative. The present state of the block, comprising the longest valid blockchain and the equity distribution gained, determines Challenge b. The following requirements apply to an unpaid transaction that is held by the node as the input:

$$Hash(a,b) \leq A_{coin} * C_{coin} \quad (10.7)$$

The current time is steadily raised in seconds, with C_{coin} representing the coin accumulation time and A_{coin} representing the coin amount. Each second, the node may make a fresh effort to see whether it has been chosen as an indicative. The ability of a node to pack is determined by whether it meets the discrepancy. Therefore the $A_{coin} * C_{coin}$ is the node's coin age, and nodes with a higher coin age have a better probability of satisfying the formula (10.7) when the difficulty level is the same.

PoS consensus protocol offers three more qualities in terms of performance: energy-saving, rapid trading speed, danger of coin age accumulation attack, and N@S attack. The following sections discuss the characteristics of the PoS protocol.

To begin with, PoS has a greater Transaction Per Second (TPS) than PoW. Because competitive packing possibilities aren't dependent on computer power, the PoS protocol is based on the stakes that nodes hold and how they vote. As a consequence, the PoS protocol reduces transaction time and increases TPS when compared to the PoW protocol.

The second point is that PoS's resilience is limited owing to two types of attacks: coin age buildup and N@S attacks.

Because of the coin age accumulation assault, the attack's resilience is minimal. The difficulty of mining in the early versions of PoS was not only tied to the current account balance, but also to the hold duration of each coin. In this instance, some nodes will eventually attain a larger C_{coin} after a time of waiting. It is simpler for larger C_{coin} to meet the formula (10.7) at the same level of coin number A_{coin} and the same difficulty d.

By raising the currency age, these nodes will be able to control the whole network. A detrimental effect on the overall system will be produced if these nodes passively pack or collude to tamper with system data.

10.3.5.2 Decryption of Data and Access Granting

Decryption is the process of restoring encrypted information to its natural position. In most cases, it's a reversal of the encryption method. Since decryption needs a secret pass code, it decrypts the encrypted files such that only legitimate personnel may decrypt the message. After decryption of data, access is granted for the end users by the blockchain network.

10.3.5.3 End Users

End users are those that will interact with the system via smart devices or a web interface. Patients, physicians, caregivers, chemists, therapists, insurance firms, pharmacist, and investigators are among those involved

10.4 PERFORMANCE ANALYSIS

The performance of the proposed system is evaluated and compared with various existing methodologies. We analyze the performance of this system using the metrics like encryption time, security level, communication cost, computation cost, and time consumption.

10.4.1 ENCRYPTION TIME

Encryption is the method of encoding data in cryptography. This procedure turns plaintext, or the original representation of the information, into cipher text, which is an alternate version. An encryption technique commonly employs a pseudo-random encryption key produced by an algorithm for operational reasons.

The encryption time is used to determine the efficiency of any encryption process, which would be computed by dividing the total encoded plaintext (in bytes) by the encryption time (in ms).

The comparative analysis of encryption time is shown in Figure 10.6. It is clear from the graph that the proposed method has a lesser encryption time.

10.4.2 SECURITY LEVEL

Medical data must be managed with privacy and safeguarded from dangerous attackers, hence security is critical in a healthcare management platform. This study explores the DO-level BEA's of security in the key segments to prove its supremacy: computational difficulty and speed tests, information entropy assessment, and resilience to differential attacks assessment, statistical attack assessment, secret key assessment, and comparison with current systems.

Design and Implementation of a Smart Healthcare System

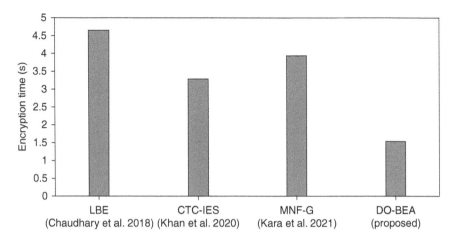

FIGURE 10.6 Comparison of encryption time.

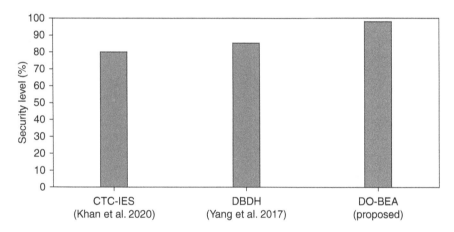

FIGURE 10.7 Comparison of security level.

The comparative analysis of security level for the existing and proposed method is shown in Figure 10.7. It is evident from the graph that the proposed method has a higher level of security when compared to the existing methodologies.

10.4.3 COMMUNICATION COST

The cost of communication comprises all overheads, which are influenced by the message's size. This covers connection bandwidth, data validation and repair, and so forth.

The comparative analysis of communication cost for the existing and proposed methods is displayed in Figure 10.8.

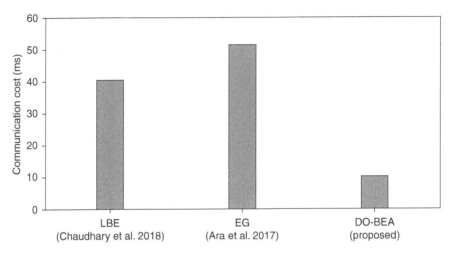

FIGURE 10.8 Comparison of communication cost.

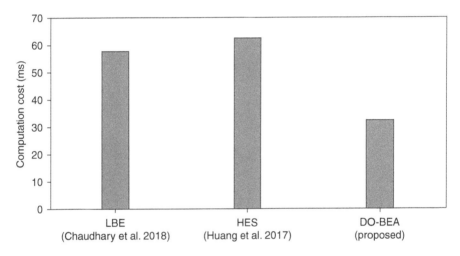

FIGURE 10.9 Comparison of computation cost.

10.4.4 Computation Cost

The computational cost of a simulation is the execution time per time step. Determine the simulation execution-time cost for the real-time target device to predict the time it would take for the simulation to run on real-time hardware. This chapter suggests cryptosystem's computational complexity and compares the cost of computing to that of competing cryptosystems.

The computational cost of healthcare applications involving complex simulations can be greatly reduced with the help of the proposed method as depicted in Figure 10.9.

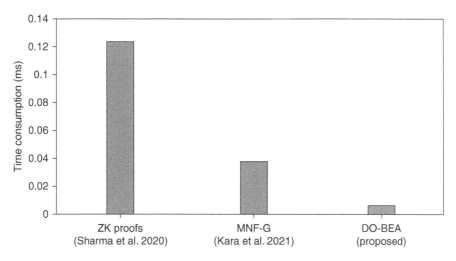

FIGURE 10.10 Comparison of time consumption.

10.4.5 TIME CONSUMPTION

The total time needed to perform the encryption and decryption of data in the blockchain system is referred to as time consumption. The comparison of time consumption is shown for the existing and the proposed systems in Figure 10.10.

10.5 CONCLUSION

Critical security threats and issues face IoT-enabled smart healthcare systems. Understanding the security needs of such systems is necessary to reduce these risks and problems. Due to the scalability, greater cost, single-point-of-failure, and resource-constrained nature of IoT devices, standard security procedures are unable to meet all of the security needs of an IoT-enabled smart healthcare system.

Blockchain has just ushered in a new age of security and anonymity in the healthcare industry. We presented a blockchain-based smart healthcare system with a unique encryption method in this study. When compared to other innovative ways, this technology has improved the security of the healthcare system.

The future of blockchain in healthcare will also need the development of infrastructure and connections in order for the technology to be widely used.

Additionally, network problems, scalability, and broad adoption of the blockchain system must be addressed. One of the greatest roadblocks to blockchain adoption is a lack of understanding among stakeholders, which must be addressed in the future (Arun Kumar Sivaraman et al., 2022).

REFERENCES

Ara, A., Al-Rodhaan, M., Tian, Y., & Al-Dhelaan, A., 2017. A secure privacy-preserving data aggregation scheme based on bilinear ElGamal cryptosystem for remote health monitoring systems. *IEEE Access*, 5, 12601–17.

Bhambri, Khang, A., Sita Rani, & Gaurav Gupta, 2022. *Cloud and Fog Computing Platforms for Internet of Things*. Chapman & Hall. ISBN: 978-1-032-101507. https://doi.org/10.1201/9781032101507.

Chakraborty, S., Aich, S., & Kim, H.C., 2019, February. A secure healthcare system design framework using blockchain technology. In *2019 21st International Conference on Advanced Communication Technology (ICACT)* (pp. 260–4). IEEE.

Chaudhary, R., Jindal, A., Aujla, G.S., Kumar, N., Das, A.K., & Saxena, N., 2018. Lscsh: Lattice-based secure cryptosystem for smart healthcare in smart cities environment. *IEEE Communications Magazine* 56(4), 24–32.

Du, X., Chen, B., Ma, M., & Zhang, Y., 2021. Research on the application of blockchain in Bhawiyuga, A., Wardhana, A., Amron, K., and Kirana, A.P., 2019, December. Platform for smart healthcare: constructing a hierarchical framework. *Journal of Healthcare Engineering, 2021*.

Farhin, F., Kaiser, M.S., & Mahmud, M., 2021. Secured smart healthcare system: blockchain and bayesian inference based approach. In *Proceedings of International Conference on Trends in Computational and Cognitive Engineering* (pp. 455–65). Springer, Singapore.

Huang, H., Gong, T., Ye, N., Wang, R., & Dou, Y., 2017. Private and secured medical data transmission and analysis for wireless sensing healthcare system. *IEEE Transactions on Industrial Informatics* 13(3), 1227–37.

Integrating Internet of Things based smart healthcare system and blockchain network. In *2019 6th NAFOSTED Conference on Information and Computer Science (NICS)* (pp. 55–60). IEEE.

Kara, M., Laouid, A., Yagoub, M.A., Euler, R., Medileh, S., Hammoudeh, M., Eleyan, A., & Bounceur, A., 2021. A fully homomorphic encryption based on magic number fragmentation and El-Gamal encryption: Smart healthcare use case. *Expert Systems* 39(5). DOI:10.1111/exsy.12767

Khan, J., Li, J.P., Ahamad, B., Parveen, S., Haq, A.U., Khan, G.A., & Sangaiah, A.K., 2020. SMSH: Secure surveillance mechanism on smart healthcare IoT system with probabilistic image encryption. *IEEE Access* 8, 15747–67.

Khang, A., Geeta Rana, Ravindra Sharma, Alok Kumar Goel, & Ashok Kumar Dubey, 2021. The role of artificial intelligence in blockchain applications, *Reinventing Manufacturing and Business Processes Through Artificial Intelligence*, 19–38, 2021, https://doi.org/10.1201/9781003145011

Khang, A., Sita Rani, Meetali Chauhan, & Aman Kataria, 2021. IoT equipped intelligent distributed framework for smart healthcare systems, *Networking and Internet Architecture*, https://arxiv.org/abs/2110.04997v2, https://doi.org/10.48550/arXiv.2110.04997

Khang, A., Rani, S., & Sivaraman, A.K. (Eds.), 2022. *AI-Centric Smart City Ecosystems: Technologies, Design and Implementation* (1st ed.). CRC Press. https://doi.org/10.1201/9781003252542

Kumar, M. & Chand, S., 2020. A secure and efficient cloud-centric internet-of-medical-thingsenabled smart healthcare system with public verifiability. *IEEE Internet of Things Journal* 7(10), 10650–9.

Miraz, M.H. & Donald, D.C., 2018, August. Application of blockchain in booking and registration systems of securities exchanges. In 2018 *International Conference on Computing, Electronics & Communications Engineering (iCCECE)* (pp. 35–40). IEEE.

Sharma, B., Halder, R., & Singh, J., 2020, January. Blockchain-based interoperable healthcare using zero-knowledge proofs and proxy re-encryption. In *2020 International Conference on COMmunication Systems & NETworkS (COMSNETS)* (pp. 1–6). IEEE.

Shukla, R.G., Agarwal, A., & Shukla, S., 2020. Blockchain-powered smart healthcare system. In *Handbook of Research on Blockchain Technology* (pp. 245–70). Academic Press.

Tripathi, G., Ahad, M.A., & Paiva, S., 2020, March. S2HS-A blockchain based approach for smart healthcare system. *Healthcare* 8(1), 100391. Elsevier.

Yang, Y., Zheng, X., & Tang, C., 2017. Lightweight distributed secure data management system for health Internet of Things. *Journal of Network and Computer Applications* 89, 26–37.

Zheng, Z., Xie, S., Dai, H.N., Chen, X., & Wang, H., 2018. Blockchain challenges and opportunities: A survey. *International Journal of Web and Grid Services* 14(4), 352–75.

11 Implementation of a Blockchain-based Smart Shopping System for Automated Bill Generation Using Smart Carts with Cryptographic Algorithms

Parin Somani, Sunil Kumar Vohra, Subrata Chowdhury, and Shashi Kant Gupta

CONTENTS

11.1 Introduction	155
11.2 Problem Statement	156
11.3 Proposed Work	157
11.3.1 Database On Server	157
11.3.2 Smart Cart	157
11.3.3 Data Validation Using Smart Contracts	159
11.3.4 Data Stored in Blockchain	160
11.3.4.1 Data Encryption Using Synchronous Discrete Twofish Encryption Algorithm	160
11.3.5 Generation of Bill	162
11.4 Results and Discussion	163
11.5 Conclusion	165
References	167

11.1 INTRODUCTION

People nowadays like to go to the mall because they can get many different items in one location. Modern shopping malls and supermarkets all include shopping carts. Recently, billions of people purchase for their daily needs at enormous amount in different shopping malls. People purchase a wide range of things and load them into their shopping carts.

When there are big deals and discounts, shopping by people rise even more. It is necessary to compute the number of things sold and prepare a bill for the consumer. Following the completion of the shopping, the product must be invoiced, and the consumer must wait in line at the billing stations, resulting in huge lines. During holidays and festivals, billing counter becomes too crowded (Aljawarneh et al., 2018).

In real time, the barcode scanner scanned the barcode tags located on every item in a supermarket as well as loyalty cards. However, issues such as store stock management and stealing are not entirely addressed by this shopping method.

In everyone's life, time is the most important commodity; no one wants to squander it. When shopping, everyone wastes the majority of their time waiting in lines at the billing counters, necessitating the addition of additional human resources to the billing department. Smart shopping cart strategies are being developed nowadays to save shopping time. Customers may construct their own invoices under this strategy, making it simple for them to estimate their cost as well (Shahnoor et al., 2021).

For automatic billing, an advanced retail system with RFID equipped smart-cart is employed. RFID readers are used to read the product tag when it is dropped into and removed from the trolley. When a product is deposited into the trolley, a reader tells the microcontroller the relevant product ID, quantity, and price.

The total amount of the bill will be computed. When the product is removed from the trolley, another RF reader transmits the product ID, which is then deleted from the bill and the revised bill amount is calculated. During both instances, the product and payment information, as well as the cart ID, are transferred to the remote PC. When the consumer completes the transaction, a final bill for the cart is created, and the bill amount is shown to the customer. After each billing session, this system also updates the product's supply (Mekruksavanich et al., 2019).

The automated billing system needs robust technology based systems for efficient storage of product data and generation of bills. Because blockchain technology is ideally adapted to the information age, it is becoming more appealing to the next generation. Blockchain technology, a distributed digital ledger that may be used to securely keep track of ever-growing lists of data records and transactions, has lately been used in shopping malls (Sharmila et al., 2021).

The most essential and distinguishing feature of the blockchain idea is that the data is completely protected inside the blocks of the blockchain's transactions. However, the blockchain is also associated with some security issues. Hence the data must be protected using encryption techniques to protect the data. Hence, this chapter applies secured blockchain technology with smart carts for automated bill generation to reduce the shopping time and manpower (Latha et al., 2021).

11.2 PROBLEM STATEMENT

Customers purchase the things and bring to the check-out area, then stand and wait in a lengthy queue so that the products may be scanned, the total amount computed, and the bill paid. This traditional method takes a long time. To avoid lengthy lines at the billing counter, save time, and shop pleasantly, effective billing management systems are required in retail establishments. Smart cart and blockchain are gaining importance in smart shopping field. Hence, this chapter proposes a blockchain based smart carts for automatic billing (Premananthan et al., 2020).

11.3 PROPOSED WORK

This chapter discusses about the blockchain based smart shopping system for automated bill generation using smart card with cryptographic algorithm. The database is stored on the server and the RFID reads the information on the reader and store it in microcontroller. Further, it will communicate with the server via Zig-Bee to request product information and display it in LCD screen which is attached to the cart displays the billing information generated by the smart cart (Rajithkumar et al., 2018).

The data is then validated and stored using smart contracts in blockchain. Then the data is encrypted using synchronous discrete twofish encryption algorithm and optimized using genetically modified firefly optimization algorithm. The bill is generated by decrypting the data and the payment is done by updating the database of the blockchain and the performance of the optimization is evaluated (Karunakara et al., 2019).

11.3.1 DATABASE ON SERVER

On a server, there is a database that includes information about all of the goods in the shop. The MySql database is being utilized for simplicity. It has two primary tables, one for product information and the other for client information. The fields in the Products table are as follows.

- RFID-enabled products (Primary key)
- The product's name
- Cost
- The product's weight

The fields of the customer table are as follows:

- Username (Primary key)
- Barcode ID for the user
- Login name
- Your password
- Previous Invoices

11.3.2 SMART CART

Ultra-high frequency passive tags are embedded in every item in the shop. Because all things can be instantly read and simply recorded, inventory management becomes simple for the shop. They recommend using ultra high frequency (UHF) RFID technology in the smart shopping system since UHF passive tags have a greater range, ranging from 1 to 12 meters.

Previous research on smart retail system design has mostly focused on low/high frequency RFID, which have limited ranges and need users to manually scan products with an RFID scanner. The suggested smart shopping system comprises of a smart cart with a Raspberry Pi-based microprocessor that communicates with an RFID reader, a Zig-Bee adapter, an LCD display, and a weight scanner mounted to the cart. The components of a smart cart are shown in Figure 11.1.

The smart shopping system is based on RFID (Radio Frequency Identification) technology, which has received little attention in the past. In such a system, all objects for sale are labelled with an RFID tag, allowing them to be monitored by any device in the shop that has an RFID reader, such as a smart shelf. Intuitively, this results in the following advantages:

Items placed in a smart shopping cart (capable of RFID scanning) can be automatically scanned, and billing information can be produced on the smart cart as well. As a consequence, consumers will not have to line for extended periods of time at the checkout.

The RFID reader reads the data recorded on tags connected to store products and sends it to the server, which stores all of the product data using a microcontroller.

Smart shelves with RFID readers can keep track of all stored products and give updates to the server about their state. When products run out, the server may send an alert to the personnel to refill.

Wireless connection between the server and the smart cart is accomplished using Zig-Bee technology. The billing information produced by the smart cart is shown on the LCD screen connected to the cart. If a hostile user rips off the RFID tags before placing them in the cart, the weight is not recorded by the weight scanner, which helps to increase security.

RFID readers are also attached to shelves to make them smart. The item's status, stock information for out-of-stock goods, customer preferences, refilling of specified items or product expiration, and other information may be communicated from the

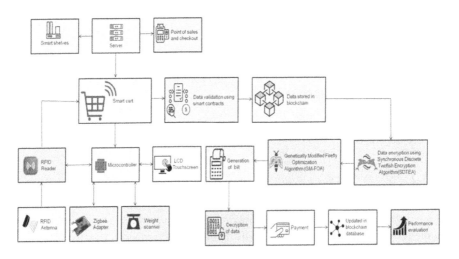

FIGURE 11.1 Flow of the proposed work.

smart shelves to the server via Zig-Bee. Before checkout, an RFID reader is deployed to guarantee that all of the things bought by the consumer have been paid for. As a result, any underpayment will set off an alarm.

All goods in the shop with the state "for sale" or "sold" are saved in the server database for this. When a product is paid for, its status in the server database is changed to "sold." As a result, only an honest consumer who has paid for all of the things purchased is permitted to leave the business.

All of the items that are put on the shelves are first registered and given RFID tags. At the server end, a database contains information such as pricing, location, and so on. When an item is placed in a smart cart, the RFID reader should scan the tag and transfer the tag information to the microcontroller, which will then interact with the server over Zig-Bee to get product information.

The smart cart is also in charge of encrypting and signing the communication. Smart cart generates two symmetric keys, s1 and s2, which are then transmitted to the server with the request. S1 is used to encrypt the information requested, while s2 is used to generate a message authentication code (MAC). The server conducts symmetric decryption and MAC testing when it gets the message.

11.3.3 Data Validation Using Smart Contracts

Nick was the first to propose the concept of smart contracts in the 1990s. However, since there was no platform to execute smart contracts prior to the introduction of blockchain technology, smart contracts were buried and failed to draw the attention of business and academics for a long period. The heyday of smart contracts, on the other hand, has already started.

The creation of smart contracts has grown in popularity, especially with the launch of Ethereum, which is based on blockchain technology. Apart from Ethereum, there are a slew of other blockchain-based systems enabling smart contract execution, including Hyperledger Fabric, Corda, Stellar, Rootstock, EOS, and others.

The blockchain and smart contracts have grown and functioned in tandem in this manner. The smart contract validates the information about the shopping item, such as its name, weight, and price. They provide safe transactions by enabling only legitimate data transfers between smart meters and supervisory nodes, and they notify any data tampering that is illegal or harmful.

Smart contracts, unlike real-world contracts, are fully digital and are simply code containers that encode. Smart contracts are computer protocols that, once built and implemented, can be self-executed and self-verified without the need for human interaction.

In a no-trust contractual context, smart contracts may help parties build confidence. When specific circumstances are met, the terms and conditions included in smart contracts will be automatically enforced.

Smart contracts, which are often put on and secured by blockchain, offer the benefits of lowering transaction risk, lowering administration and service costs, and boosting business process efficiency when compared to conventional contracts. Smart contracts are projected to give a superior solution to the present transaction mechanism in a variety of businesses in this regard.

11.3.4 DATA STORED IN BLOCKCHAIN

The shopping item data and the customer details such as name of the product, weight of the product, price, user ID, user barcode ID, username, password and the past bills are stored in the blockchain.

11.3.4.1 Data Encryption Using Synchronous Discrete Twofish Encryption Algorithm

The synchronous discrete Twofish encryption technique, like the symmetrical block cypher, has a fiestel structure. It is particularly useful for programs that operate on tiny processors (intelligent cards) or for hardware embedding. It provides users with the ability to optimize encryption speed, key configuration, and code size in order to achieve performance equilibrium.

The synchronous discrete Twofish encryption method is completely free of charge, unpatented, and available for download. Double-fish encryption is used with 128 bits, 192 bits, and 256 bits of key length. In this case, the block size is 128 bits, and the encryption process is made up of 16 loops.

Among the primary building blocks of Feistel Networks is the F work, which consists of the following elements:

- A key-subordinate mapping of an information string onto a yield string;
- An F work is reliably non-direct and possibly non-surjective;

$$U : \{0,1\}^{k/2} \times \{0,1\}^{k} \mapsto \{0,1\}^{k/2} \quad (11.1)$$

Where n is the Feistel Network's piece size, and U is a capacity that delivers a yield of length k /2 bits by using k /2 bits of the square and K bits of a key as information.

11.3.4.1.1 Synchronous Discrete Twofish Encryption Algorithm
 1. User key "K" is determined.
 2. Sub-key generation function is invoked for generating sub-keys.
 3. The encryption function U is invoked to generate the round keys. The random input is sent as input to this function.
 4. Step-3 is repeated to obtain all round keys,
 5. Block "B" is read from the given data.
 6. The block is divided into Left and Right sub-blocks.
 7. The left block is encrypted by invoking the encryption function.
 8. The right block is modified by merging the right block with the resultant from step 7 using XOR function.
 9. Left and right blocks are swapped.
 10. Steps 7 to 9 are repeated for 16 rounds.
 11. After 16 rounds, the resultant blocks are combined into one block.
 12. If data is not finished.
 13. The next block of the data is loaded into the encryption process.
 14. Steps from 6 to 13 are repeated.

15. If the data is finished then encryption process is finished.
16. Encrypted data is obtained.

11.3.4.1.2 Genetically Modified Firefly Optimization Algorithm

The original genetically modified Firefly Algorithm was developed. It's based on the way fireflies interact with one another. Short, rhythmic flashes are produced by the majority of genetically engineered fireflies to entice mates and possible prey. Using the following three principles, the genetically modified firefly optimization method is described:

1. Fireflies that have been genetically engineered to be unisex are now available. A genetically engineered firefly optimization will be drawn to another genetically modified firefly optimization no matter what their gender.
2. Their brightness is directly related to their attractiveness. Because of this, if there are two genetically modified fireflies that are both flashing, the brighter one will attract the less brilliant one. The attraction is inversely proportional to the distance from the object. As long as there are no additional genetically modified fireflies that are brighter than the current brightest genetically modified firefly in its present form, it will travel in an erratic fashion.

The landscape of the goal function affects or determines the brightness of a genetically engineered firefly. The brightness may be inversely proportional to the objective function in a minimization problem.

An improved genetically engineered firefly will have a lower objective function value if it is brighter. When it comes to a genetically engineered firefly, light intensity has a big role. The following equation may be used to estimate attractiveness:

$$\beta(n) = \beta_0 v^{-\gamma n^2} \tag{11.2}$$

Where $\beta(n)$ is the attractiveness at n = 0, and n is the distance between two genetically modified fireflies. The parameter γ is the light absorption coefficient, which is usually set to 1.

For two genetically modified fireflies A_i and A_j, their distance n_{ij} can be defined by

$$n_{jp} = \|A_j - A_p\| = \sqrt{\sum_{o=1}^{O}(a_{jo} - a_{po})^2} \tag{11.3}$$

Where O is the problem dimension. The movement of a genetically modified firefly A_i, which is attracted to another brighter firefly A_j, is determined by

$$a_{jo}(i+1) = a_{jo}(i) + \beta_0 v^{-\gamma n_{jp}^2}(a_{po}(i) - a_{jo}(i)) + \alpha\varepsilon \tag{11.4}$$

Where a_{po} and a_{jo} are the d-th dimension value of firefly A_i and A_j, respectively. The step parameter α is a random value within the range [0, 1], ε is a Gaussian random number for the d-th dimension, and i indicates the index of generation.

Genetically Modified Firefly Algorithm
Inputs: population size "n," maximum generation "MG"

1. Alpha, beta and gamma values are initialized
2. Initial fitness value of "n" fireflies are computed
3. The initial best solution is determined
4. For every iteration/generation
5. For all fireflies
    a. New solution is computed
    b. All fireflies are sorted depending on their light intensities iii. Current best solution is estimated
6. end for loop
7. For all "n" fireflies with variable i
    a. For all "n" fireflies with variable j if I(i)>I(j)
        i. = + βexp[-γ] (-) + αt εt
        ii. Where β=β0
        iii. End of if loop
        iv. New solutions are determined and light intensity is updated End of for loop
    b. End of for loop
8. End of for loop
9. The best solution is printed.

11.3.5 Generation of Bill

Three algorithms are used to complete the billing generation process. Current system is denoted by T.

Algorithm 1: When the smart cart scans the Product RFID tag, it validates the HMAC contained on the tag. Assuming the verification was successful, the Smart cart produces s1 and s2 at random. To encrypt a communication, S1 is utilized, whereas S2 is used to generate the message authentication code. Signs and encrypts tag information with its Own ID and time stamp, and transmits encrypted message to server using session keys S1 and S2.

Algorithm 2, the server decrypts the message and checks the signature and time stamp to ensure the message is authentic. To verify that a message is genuine, a server searches the database for the item's requested information (TI) and adds it to the message along with the updated time stamp, then encrypts it using S1 acquired from the shopping cart. S2 is also used to generate the Message authentication code, which is sent to the smart Cart along with the encrypted message.

Algorithm 3: The smart card initially validates the MAC using s2 after getting the response from the Server. The smart card decrypts the message if the MAC is valid and Checks the validity of the time stamp using s1.

After verification, the smart cart will show the Billing information on the LCD. This safe and intelligent system's most important task is to provide accurate item readings. A client may place an item in their basket, only to subsequently remove it and return it. This modification should be picked up by the cart's built-in sensors. So, it should be able to distinguish between products that have been added to or withdrawn from the cart.

In addition, a cart should not be able to read the contents of another cart that is close by. In order to prevent dishonest customers from attempting to leave the shop without paying, payment verification is essential. The secrecy and integrity of the system are preserved because fresh session keys are produced for each session and utilized by the smart cart and server for communication.

The time stamp T is resistant to replay assaults because of its date and time. Tag switching is also not possible due to the fact that they are tamper-proof, and an HMAC on the tag is confirmed before the bill is created. To put it another way, this increases the safety of the smart purchasing system.

This safe and intelligent system's most important task is to provide accurate item readings. A client may place an item in their basket, only to subsequently remove it and return it. This modification should be picked up by the cart's built-in sensors. So, it should be able to distinguish between products that have been added to or withdrawn from the cart.

In addition, a cart should not be able to read the contents of another cart that is close by. In order to prevent dishonest customers from attempting to leave the shop without paying, payment verification is essential. The secrecy and integrity of the system are preserved because fresh session keys are produced for each session and utilized by the smart cart and server for communication.

The time stamp T is resistant to replay assaults because of its date and time. Tag switching is also not possible due to the fact that they are tamper-proof, and an HMAC on the tag is confirmed before the bill is created. As a result, the smart shopping system is safe and reliable. The decrypted data is then updated in the database of blockchain and the payment is made by the customer.

11.4 RESULTS AND DISCUSSION

The performance of blockchain based smart carts for automatic billing was analyzed in this section. Smart cart data was stored in blockchain and are encrypted using SDTEA and the encryption efficiency was optimized by applying GM-FOA. The performance of blockchain based smart carts was compared with the cloud based smart carts and traditional approach which is manual billing system to evaluate its application in automatic billing (Bhambri et al., Cloud & IoT, 2022).

From Figure 11.2, it is clear that time taken for shopping using the proposed system was less compared to manual billing and cloud based smart carts. It is observed that there is a significant reduction in shopping and billing time when blockchain based smart cart was applied in shopping malls.

This result ensures that the proposed approach is an efficient shopping method for people. The approximate comparison of the shopping time between the traditional technique and the suggested system is shown in Table 11.1.

Table 11.1 displays comparative analysis of performance of proposed technique with existing techniques

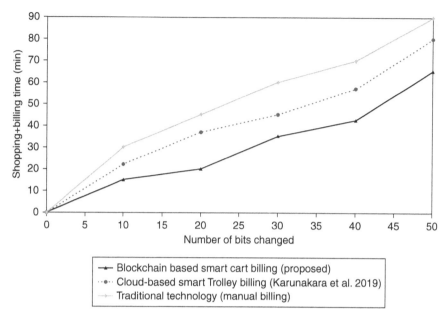

FIGURE 11.2 Comparison of blockchain based smart cart billing with existing techniques.

TABLE 11.1
The Traditional Technique and the Suggested System

	Time required for shopping and billing (minutes)		
Number of items purchased	Blockchain based smart cart billing (proposed)	Cloud based smart cart billing	Traditional technique (Manual billing)
10	15	22	30
42	20	37	45
30	35	45	60
40	42	57	70
50	65	80	90

The proposed encryption (SDTEA+GM-FOA) technique was compared to the existing methods such as Advanced Encryption Standard (AES) and simple and highly secure encryption decryption (SHSED) algorithm based on the performance metrics namely encryption time, decryption time, and avalanche effect. This analysis was carried out to prove the efficiency of encryption technique in blockchain.

Encryption time is the time taken to complete the encryption process that is converting the original data into encrypted data. Its unit is seconds. From Figure 11.3, it is evident that SDTEA+GM-FOA takes lesser encryption time compared to AES and SHSED. Decryption time is the time taken to complete the decryption process that is converting the encrypted data into original data.

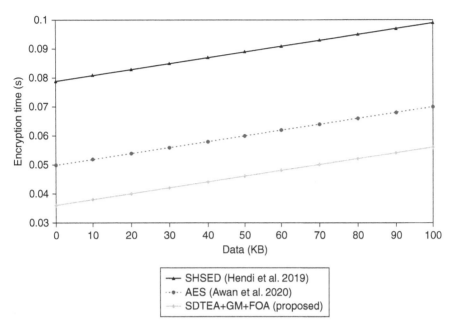

FIGURE 11.3 SDTEA+GM-FOA.

From Figure 11.4, it is evident that SDTEA+GM-FOA takes lesser decryption time compared to AES and SHSED. Lesser time for encrypting and decrypting the data ensures that the suggested approach executes the cryptographic processes very fastly.

The avalanche effect is one of the important security metrics for validating the encryption algorithm. It should be mentioned that the Avalanche Effect's key benefit was establishing the robustness of our algorithm against cracking and hacking threats.

Avalanche effects are determined using the equation 11.5. If an input is changed slightly, the output must change significantly.

$$\text{Avalanche effect} = \frac{\text{Number of bits changed in the encrypted data}}{\text{Total number of bits encrypted data}} \times 100 \quad (11.5)$$

From Figure 11.5, it is noted that the proposed technique achieved higher avalanche effect compared to AES and SHSED. This denotes that higher number of bits is modified in the encrypted data when there is a slight change in the input data. This means that SDTEA+GMFOA have a higher level of security than AES and SHSED.

11.5 CONCLUSION

Customers would benefit greatly from the use of smart shopping carts since they will be able to avoid the inconveniences that they regularly face when shopping. In this

166 The Data-Driven Blockchain Ecosystem

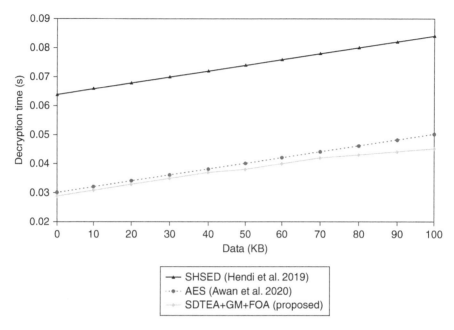

FIGURE 11.4 Proposed encryption method.

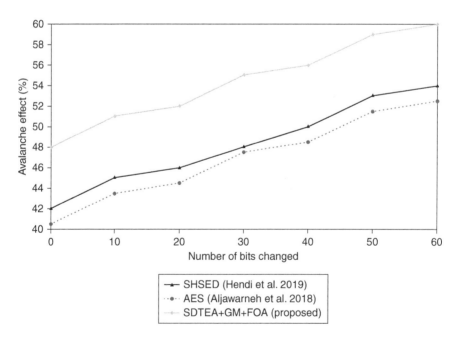

FIGURE 11.5 SDTEA+GMFOA.

chapter, the blockchain technology was applied along with smart carts for faster and secured automatic bill generation.

Shopping data and calculated amount were encrypted using SDTEA and the encryption process was optimized using GM-FOA. The result analysis shows that the proposed encryption technique (SDTA+GM-FOA) achieved lesser execution time and higher avalanche effect compared to AES and SHSED.

When blockchain-based smart carts are utilized instead of traditional shopping techniques, shopping time is reduced. Hence, the automatic billing system dependent on smart cart with blockchain can tackle the problem of long waiting queues, higher man power and shopping time (Khang et al., 2022).

REFERENCES

Aljawarneh, S., Yassein, M.B., & Talafha, W.A.A. 2018. A multithreaded programming approach for multimedia big data: encryption system. *Multimedia Tools and Applications* 77(9), 10997–11016.

Arora, J., Gagandeep, S.J., & Kumar, R., 2018. Smart goods billing management and payment system for shopping malls. *International Journal of Engineering & Technology* 7(2.7): 456461.

Awan, I.A., Shiraz, M., Hashmi, M.U., Shaheen, Q., Akhtar, R., & Ditta, A., 2020. *Secure Framework Enhancing AES Algorithm in Cloud Computing*. Security and Communication Networks, 2020.

Bhambri, Khang, A., Sita Rani, Gaurav Gupta, 2022. *Cloud and Fog Computing Platforms for Internet of Things*. Chapman & Hall. ISBN: 978-1-032-101507, https://doi.org/10.1201/9781032101507

Karunakara Rai, B., Harshitha, J.P., Kalagudi, R.S., Chowdary, P., Hora, P., & Sahana, B., 2019. A cloud-based inventory management system using a smart trolley for automated billing and theft detection. In *Innovations in Electronics and Communication Engineering* (pp. 491–500). Springer, Singapore.

Khang, A., Sita Rani, Meetali Chauhan, & Aman Kataria, 2021. IoT equipped intelligent distributed framework for smart healthcare systems, *Networking and Internet Architecture*, https://arxiv.org/abs/2110.04997v2, https://doi.org/10.48550/arXiv.2110.04997

Khang, A., Geeta Rana, Ravindra Sharma, Alok Kumar Goel, & Ashok Kumar Dubey, 2021. The role of artificial intelligence in blockchain applications, *Reinventing Manufacturing and Business Processes through Artificial Intelligence*, 19–38, https://doi.org/10.1201/9781003145011

Khang, A., Vladimir Hahanov, Gardashova Latafat Abbas, & Vugar Abdullayev Hajimahmud, 2022. Cyber-physical-social system and incident management, *AI-Centric Smart City Ecosystems: Technologies, Design and Implementation* (1st ed.). CRC Press. https://doi.org/10.1201/9781003252542

Khang, A., Rani, S., & Sivaraman, A.K. (Eds.), 2022. *AI-Centric Smart City Ecosystems: Technologies, Design and Implementation* (1st ed.). CRC Press. https://doi.org/10.1201/9781003252542

Latha, G.C.P., Kandhasamy, J.P., & Sridhar, S., 2021. Smart shopping cart by RFID technology. *Materials Today: Proceedings*.

Mekruksavanich, S., 2019, October. The smart shopping basket based on iot applications. In *2019 IEEE 10th International Conference on Software Engineering and Service Science (ICSESS)* (pp. 714–17). IEEE.

Premananthan, G., Nagaraj, B., & Divya, N., 2020. Sensor based integrated smart trolley system using Zigbee-experimental analysis. *Materials Today: Proceedings*.

Rajithkumar, B.K., Deepak, G.M., Uma, B.V., Hadimani, B.N., Darshan, A.R. & Kamble, C.R., 2018. Design and development of weight sensors based smart shopping cart and rack system for shopping malls. *Materials Today: Proceedings*, 5(4), 10814–20.

Shahnoor, S.F., 2021. Smart cart for automatic billing with integrated RFID system. *Turkish Journal of Computer and Mathematics Education (TURCOMAT)*, 12(12), 2487–93.

Sharmila, G., Ragaventhiran, J., Islabudeen, M., & Kumar, B.M., 2021. RFID based smart-cart system with automated billing and assistance for visually impaired. *Materials Today: Proceedings*.

12 Multi-Node Data Privacy Audit for Blockchain Integrity

A. Shenbaga Bharatha Priya, Sanjaya Kumar Sarangi, S. Balasubramanian, and Bhaskar Roy

CONTENTS

- 12.1 Introduction 169
- 12.2 Problem Description and Analysis 171
  - 12.2.1 Problem Description 171
  - 12.2.2 Probability Analysis of Cross-Shard Transactions 173
  - 12.2.3 Probability of Cross-Shard Transaction Rollback 174
- 12.3 Multi-Round Consensus Processing Cross-Shard Transaction Verification Scheme 176
  - 12.3.1 The Process of Processing Transactions in the Multi-Round Verification Scheme 176
  - 12.3.2 Selection of the Number of Rounds of the Multi-Round Verification Scheme 177
    - 12.3.2.1 Probability of Byzantine Collision Attack 177
    - 12.3.2.2 Influence of Multiple Rounds on Rollback Probability 178
    - 12.3.2.3 The Upper Limit of the Number of Rounds 179
  - 12.3.3 Node Random Allocation Algorithm 180
- 12.4 Experimental Setup and Result Analysis 181
  - 12.4.1 Experimental Environment and Parameter Settings 181
  - 12.4.2 Experiment Design and Result Analysis 182
    - 12.4.2.1 Transaction Verification Rate Test 182
    - 12.4.2.2 Transaction Throughput Test 183
- 12.5 Conclusion 185
- References 185

12.1 INTRODUCTION

Blockchain technology was first proposed in Satoshi Nakamoto's paper (Chen et al., 2018). At the technical level, blockchain is a distributed digital ledger that uses a

P2P network, distributed storage, cryptography and other computer technologies and has the characteristics of decentralization, trustlessness, and difficulty to tamper (Bhambri et al., Cloud & IoT, 2022).

Due to its technical aspects, the application of blockchain has been extended to many fields such as banking, logistics, and finance, which has a significant impact on their data storage and sharing (Gräther et al., 2018).

At the same time, at the 18th joint study meeting of the Political Bureau of the Central Committee on October 24, 2019, General Secretary Xi Jinping emphasized that it is necessary to vigorously develop key blockchain technologies and accelerate the development of blockchain technology and industrial innovation (Rivera-Vargas et al., 2019).

The technical characteristics of blockchain give it broad application prospects, but it also faces the bottleneck of insufficient scalability, and there is a need for expansion (Kosba et al., 2016; Wang et al., 2020, 2021; Emilie et al., 2016).

Existing capacity expansion technologies mainly include sharding schemes (Sharples et al., 2016; Mao et al., 2021), DAG (Sun et al., 2018; Zhang et al., 2021), block expansion (Redman et al., 2016; Liu et al., 2020), side-chain technology (Williams et al., 2019; Liu et al., 2021), state channel (Yang et al., 2017), etc. Among them, the sharding scheme is the most feasible scheme among the current expansion schemes and has attracted the attention and attention of scholars and community personnel.

The core idea of blockchain sharding is to divide the blockchain network nodes into several sets (Serranito et al., 2020; Mizrahi et al., 2020); each set runs the consensus protocol independently, performs consensus verification on transactions, and each set can process different transaction sets in parallel or even only store part of the network state, to achieve the effect of improving transaction throughput.

Although sharding technology can improve the performance of blockchain systems, it also brings new challenges. After sharding, there must be cross-shard transactions in the network, and the correct processing of cross-shard transactions is critical to the system's performance.

Under the UTXO (unspent transaction outputs, unspent transaction output) model, the client sends the cross-shard transaction to the involved input shards for independent verification processing when processing a cross-shard transaction. When all input shards are successfully verified, they will be submitted to the output shard for verification processing after the cross-shard transaction.

After any input shard verification transaction fails or times out, to ensure the consistency and integrity of the blockchain network data, it is necessary to roll back the partially processed cross-shard transaction.

The reasons for the rollback of a cross-shard transaction are as follows:

a) The transaction is invalid, some input shards are verified and submitted, and a particular input shard verification transaction is null. The transaction is rejected, and the cross-shard transaction must be rolled back as a result. Release the locked resources of other shards to ensure the regular use of the object in subsequent transactions;

b) For the sharding scheme of the PBFT (Zhong et al., 2017) type of consensus algorithm adopted by most sharding projects; there is a proportion of total

Byzantine nodes. No more than 1/3, and the balance of Byzantine nodes in a single shard after sharding is more significant than 1/3, which threatens the validity of shard verification (Gresch et al., 2019; Sohrabi et al., 2020), and the cross-shard transaction allocated to a shard cannot reach a consensus. It causes cross-shard transactions to be rolled back.

Given the problem of cross-shard transaction rollback due to fragmentation failure, Vidal et al. (2019) adopted the random allocation algorithm of Rand Hound nodes to reduce the probability of aggregation of Byzantine nodes within a fragment and improve the validity of fragmentation verification. Ensure correct processing of cross-shard transactions.

San et al. (2019) increases the tolerance of a single shard to Byzantine nodes by setting the communication mode of nodes to be synchronous and improves the validity of shard verification, thereby reducing the occurrence of cross-shard transaction rollback. However, the communication method application has limitations.

Kirilova et al. (2018) performs sharding through smart contracts and uses the "SMART's PBFT" protocol assuming intra-shard security to ensure the processing of cross-shard transactions.

Bucea-Manea-Țoniş et al. (2021) proposes an incentive scheme for continuous crossbow mining. Under the economic model of improving miners' income, ensuring the effectiveness of sharding can reduce the occurrence of cross-shard transaction rollbacks to a certain extent.

According to the above literature, existing sharding schemes lack "span segments." In a shard transaction, the transaction in the shard cannot be verified because the shard is invalid. And lead to "cross-shard transaction rollback" probability analysis, as well as different processing scenarios for cross-shard transaction rollback. Therefore, after analyzing the above problems, an improved scheme for multi-round consensus verification is proposed.

The more traditional single-round verification contribution is as follows:

a) This scheme can replace the verification nodes of invalid shards in time so that the transactions in the invalid shards can be processed again in the next round of verification, reducing the effective cross-sharding. The probability of transaction rollback reduces the resubmission of the rollback transaction, resulting in a more significant delay.
b) The scheme has passed multiple rounds of verification, which can increase the scale of shards based on reducing the probability of cross-shard transaction rollback. Further, improve the TPS of the system;
c) Experiments show that the multi-round verification case performs better.

12.2 PROBLEM DESCRIPTION AND ANALYSIS

12.2.1 Problem Description

The UTXO model has been used in many applications due to its good concurrency in a multi-blockchain network. In the UTXO model, a transaction can have multiple inputs and outputs.

During the transaction verification process, the validate needs to verify that the information has not been spent and ensure that any inputs to the transaction are no longer available for spending. Shading area based on UTXO model in a blockchain system, multiple transaction inputs and outputs may involve multiple node addresses.

According to the shading protocol, nodes and transactions are divided into different shards, which lead to some transactions that need to be Verification is done on multiple chips, and this is a cross-shard transaction.

The partition that manages the input object is typically set in a cross-shard transaction. A shard is called an input shard (IS), and it contains the shard in which the output objects reside. Shards are called output shards (OS). Since the cross-shard transaction involves multiple shards, the processing between the shards is more coordinated.

When processing cross-shard transactions, the client sends shard transactions to the input shards applied, each shard independently of commerce in this shard are processed for consensus verification, and only all input shards cross shards Transactions are committed to one or more output shards and revalidated in the output shards only after shard validation is successful.

When any input fails to verify the transaction or exceeds the time limit, to ensure the consistency of the blockchain system (Ocheja et al., 2018), it is necessary to roll back the partially processed cross-shard transaction as Figure 12.1.

The sharding scheme using the PBFT consensus algorithm can ensure the final consistency of the data. Still, in this sharding scheme, there is a problem that the validity of the sharding verification is destroyed due to the uneven distribution of Byzantine nodes. Furthermore, it affects the processing of transactions within a shard.

Once a chip fails, the cross-shard transaction involving the fragment cannot be verified. The cross-shard marketing submitted by other bits involved in the cross-shard transaction needs to start the rollback operation, and rollback is critical to performance. The negative impact is enormous. Let the total number of nodes be N; the

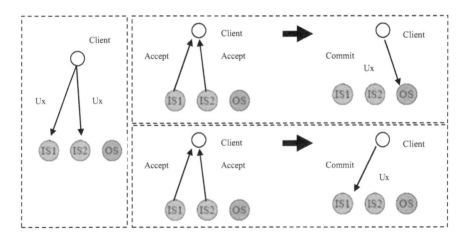

FIGURE 12.1 Processing of cross-shard transactions.

Multi-Node Data Privacy Audit for Blockchain Integrity

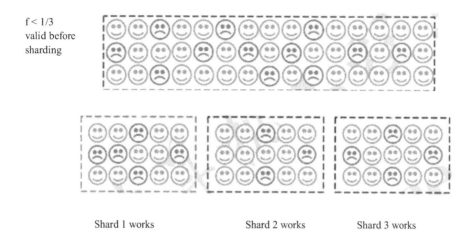

FIGURE 12.2 Uneven distribution of Byzantine nodes.

overall Byzantine ratio is f < 1/3, and the N nodes are randomly and evenly divided into k shards.

The number of nodes in the chip is L; there may be a single shard after sharding. The slice Byzantine ratio fi > 1/3 ($1 \leq i \leq k$) causes sharding to fail, as shown in Figure 12.2.

For an effective cross-shard transaction, because the proportion of Byzantine nodes in a particular shard is f > 1/3, it is impossible to reach consensus verification for the intra-shard transaction.

The resulting cross-shard transaction rollback will seriously affect the system's performance, and subsequent users. Re-sending the transaction to the network for re-validation will result in a more considerable transaction delay. Therefore, this chapter analyzes and handles the cross-shard transaction rollback problem caused by too many Byzantine sections within a shard.

12.2.2 Probability Analysis of Cross-Shard Transactions

In the sharding system of the blockchain UTXO model, multiple inputs of a transaction may involve multiple node addresses, and the nodes will be randomly allocated to each shard according to the random allocation algorithm in the sharding protocol.

For example, let the shard size be k, and the number of transaction inputs is m. According to the random allocation algorithm of shards, each input node is randomly assigned to k shards with equal probability. Then in the sharding system under the UTXO model, the transaction with m inputs is the probability P (cross-shard) of cross-shard marketing as shown in formula (12.1):

$$Q(Cross-Shard) = 1 - \left(\frac{L}{1}\right)\left(\frac{1}{L}\right)^n \qquad (12.1)$$

TABLE 12.1
Impact on Cross-Shard Transaction Probability Q (Cross-Shard)

L	n = 2	n = 4	n = 8	n = 16
1	0	0	0	0
2	0.5	0.875	0.992	0.999
4	0.75	0.984	0.999	0.999
8	0.875	0.998	0.999	0.999
16	0.938	0.999	0.999	1
32	0.969	0.999	0.999	1

According to formula (1), the influence of the input number m of the research transaction on the cross-shard transaction probability Q (cross-shard) is simulated. As shown in Table 12.1.

It can be seen from Table 12.1 that the probability that a transaction with multiple inputs is a cross shard transaction increases with the increase of the shard size k and the number of inputs m. For example, when the shard size k is 16, the probability of a transaction with multiple inputs being a cross-shard transaction exceeds 93%; if k and m continue to increase, the likelihood of a cross shard transaction will be infinitely close to 1.

To sum up, in a sharded system, when dealing with transactions with multiple inputs, there is a high probability of cross-shard transactions. Therefore, ensuring the correct processing of cross-shard transactions involving numerous inputs and reducing the likelihood of cross-shard transaction rollbacks are critical to the performance of the blockchain system.

12.2.3 PROBABILITY OF CROSS-SHARD TRANSACTION ROLLBACK

Let the shard size be k. In the UTXO model, the number of inputs m contained in a transaction may be greater than the number of shards k, then the cross-shard distribution w of a transaction U_x having m inputs takes a value of 1 to min (m,k) shards.

To focus on the analysis and comparison of cross-shard transaction rollback, a transaction U_x with m inputs in the sharding system, the probability of cross-shard transaction rollback is reduced to the cross-shard distribution w and the verification validity of the shards related.

Let the size of the bit be k, Qr represent the probability of a single shard verification failure, and the transaction U_x with m inputs distributed in w shards, its rollback probability is denoted by Q(rollback), as shown in formula (12.2):

$$Q(rollback) = 1 - (1 - Q_R)^\omega \qquad (12.2)$$

In Vidal et al. (2019), the Omniledger scheme conducts a specific analysis of the effectiveness of sharding. Where $Y = \sum_{i=1}^{M}$, Yi indicates that the ith node exhibits

Byzantine properties, Y shows the number of Byzantine nodes in a single shard, M denotes the number of nodes in the bit, and g is the overall Byzantine node proportion. Y conforms to the Binomial (M,g,Y) binomial distribution, and when $Y \geq 1/3$, the sharding fails.

The fragmentation failure probability Qs is shown in formula (12.3):

$$Q_R = Q\left(Y \geq \frac{M}{3}\right) = \sum_{y=\left|\frac{M}{3}\right|}^{M} Q_r[Y=y] = \sum_{y=\left|\frac{M}{3}\right|}^{M} \binom{M}{y} g^x (1-g)^{(M-x)} \quad (12.3)$$

Therefore, Q (rollback) can be expanded as shown in formula (12.4):

$$Q(rollback) = 1 - \left(1 - \sum_{y=\left|\frac{M}{3}\right|}^{M} \binom{M}{y} g^x (1-g)^{M-x}\right)^{\omega} \quad (12.4)$$

Setting the overall Byzantine ratio g = 0.25 [6, 16-18] in many sharding projects is considered a reasonable value. Therefore, when g = 0.25 is set, the effects of different w on the rollback probability of cross-shard transactions are calculated, as shown in Table 12.2.

According to Table 12.2, when M is constant, the probability of rollback of a cross-shard transaction increases with the increase of w; when the number of input shards w of a cross-shard transaction is stable, the likelihood of rollback rises with the rise of w. It decreases with the increase in the number of nodes M in the shard.

But even when M = 600, the probability of cross-shard transaction rollback cannot be ignored.

It can be seen from Table 12.2 that by increasing M, the rollback probability of the same cross-shard transaction can be reduced. Still, the purpose of the sharding technology is to increase the throughput by expanding the scale of the shard. Unfortunately, too many nodes in the chip are not conducive to the partition.

As a result, the scale of the bit is expanded so that the final purpose of sharding cannot be achieved. Therefore, this chapter proposes a consensus algorithm for multi-round verification, which can boost the scale of shards and improve the overall performance of the system based on reducing the probability of cross-shard transaction rollback.

TABLE 12.2
Influence of the Number of Input Shards "w" on the Rollback Probability Q (Rollback)

w	M = 60	M = 120	M = 200	M = 300	M = 400	M = 600
1	0.092	0.025	0.006 7	7.50E-04	1.30E-04	2.90E-06
2	0.18	0.05	0.014 0	1.50E-03	2.70E-04	5.90E-06
4	0.32	0.097	0.027 0	3.00E-03	5.30E-04	1.20E-05
8	0.54	0.18	0.053 0	5.90E-03	1.10E-03	2.40E-05
16	0.79	0.34	0.100 0	1.20E-02	2.20E-03	4.70E-05
32	0.96	0.54	0.200 0	2.40E-02	4.30E-03	9.50E-05

12.3 MULTI-ROUND CONSENSUS PROCESSING CROSS-SHARD TRANSACTION VERIFICATION SCHEME

It can be seen from the above analysis that there will be not only cross-shard transactions after sharding, but also cross-shard transactions will account for a large proportion of all transactions in the network. Therefore, the correct processing of cross-shard transactions is critical to system performance. Important.

When processing cross-shard transactions, even if only one input shard has not completed the verification transaction, other processed cross-shard transactions need to be rolled back. The simulation calculation in Section 13.3 analyzed the probability of cross-shard transaction rollback caused by the uneven distribution of Byzantine nodes after sharding, which caused some shards to fail. We found that the rollback probability of cross-shard transactions was very high. Therefore, it cannot be ignored.

The rollback of cross-shard transactions will seriously affect system performance. Thus, the rollback probability of cross-shard transactions should be minimized to ensure system performance. Therefore, a multi-round consensus verification scheme is proposed. Through multi-round verification, the verification rate of cross-shard transactions can be improved, thereby ensuring the overall TPS of the system.

12.3.1 THE PROCESS OF PROCESSING TRANSACTIONS IN THE MULTI-ROUND VERIFICATION SCHEME

The multi-round consensus verification scheme verifies the cross-shard transaction. The central idea is that when a cross-shard transaction is sent to multiple input shards, each shard independently performs verification processing. If the verification is successful, a block is packaged. For example, suppose there is a transaction in the bit due to too many Byzant nodes and the transaction verification times out.

In that case, the nodes will be reassigned to the shard. The transaction will be verified by consensus again to be successfully verified within a limited number of rounds and try to avoid cross-border verification.

A sharded transaction cannot complete valid validation due to the invalidation of a single input shard. This reduces the occurrence of cross-shard transaction rollbacks and also avoids more significant delays in resending rolled-back transactions.

The specific process of transaction verification is:

a) Allocate the transaction to a given shard according to the mapping rules and initialize the transaction round;
b) All nodes in the shard perform consensus verification on all transactions allocated to the shard, If the verification is successful, the group of transactions will be packaged into the block;
c) If a valid consensus has not been obtained after the shard verification timeout, a group of nodes will be reassigned to the failed shard through a random allocation algorithm (the other shards are regular at this time).

If no good consensus has been reached after consecutive T rounds, choose to abandon the transaction. The limited sacrifice of transaction availability preserves the system's overall performance; this chapter sets the probability of abandoning a transaction as $\beta < 10-8$.

12.3.2 Selection of the Number of Rounds of the Multi-Round Verification Scheme

12.3.2.1 Probability of Byzantine Collision Attack

Even if the proportion of the overall Byzantine nodes is less than 1/3, after the nodes are randomly assigned to each shard, there will be cases where the Byzantine nodes in the bit occupy the majority and collude.

The purpose of colluding malicious nodes is to conduct collusive attacks and achieve consensus and wrong consensus within the shard. If a collision is reached in the bit, the proportion of the joint evil Byzantine nodes in the bit needs to be greater than 2/3, but this has great complexity. Assume that as long as the Byzantine nodes in the chip are more significant than (2M)/3 + 1, the Byzantine nodes will collide to achieve a collision attack.

In Sharples et al. (2016), the Elastico scheme is proposed, and the collision probability of sharding is analyzed. Where $Y=\sum_{i=1}^{M}$, Yi indicates that the ith node exhibits Byzantine attributes, Y shows the number of Byzantine nodes in a single shard, L is the number of nodes in the bit, and f indicates the proportion of the total number of Byzantine nodes. Y conforms to the binomial distribution of Binomial (M,g,Y), and the probability of intra-shard collusion attacks is shown in formula (12.5):

$$Q\left(Y \geq \frac{M}{3}\right) = \sum_{y=\left\lfloor\frac{2M}{3}\right\rfloor+1}^{M} Q_r[Y=y] = \sum_{y=\left\lfloor\frac{2M}{3}\right\rfloor+1}^{M} \binom{M}{y} g^x (1-g)^{(M-x)} \quad (12.5)$$

According to formula (12.5), the collusion probability under different M and g is simulated and calculated, as shown in Table 12.3.

According to Table 12.3, when the number of nodes in a single shard is more significant, the probability of a collision attack on the chip is lower. For example, it can be seen that when g = 0.333, as long as the number of nodes M in the shard exceeds 60, the probability of collusion attack is lower than 10–8, the likelihood of collision

TABLE 12.3
The Influence of the Number of Nodes "M" and the Byzantine Ratio "g" on the Collusion Probability Q

M	g = 0.15	g = 0.20	g = 0.25	g = 0.333
15	6.54E-07	1.25E-05	1.15E-04	1.79E-03
30	1.78E-11	4.48E-09	2.82E-07	4.36E-05
45	5.34E-16	1.77E-12	7.53E-10	1.15E-06
60	1.68E-20	7.30E-16	2.10E-12	3.15E-08
90	1.78E-29	1.34E-22	1.75E-17	2.52E-11
180	2.72E-56	1.03E-42	1.28E-32	1.63E-20

is low enough, and the decision reached by the nodes in the shard is approximately considered to be collective honesty, regardless of cooperation the occurrence of an attack.

12.3.2.2 Influence of Multiple Rounds on Rollback Probability

According to Table 12.3, when L is not less than 60, the probability of collusion attack is small enough, and it can be assumed that the decision reached by the nodes in the shard is collective honesty. Therefore, Table 12.4 analyzes the rollback probability of cross-shard transactions under different rounds U and other input shards w when the number of shard nodes is M = 60.

As shown from Table 12.4, compared with the traditional single-round verification scheme (probability of rollback in the first round), the verification scheme using multiple games of consensus significantly reduces the rollback probability of cross-shard transactions. As the number of rounds T increases, the likelihood of cross-shard transaction rollback will become lower and lower, gradually decreasing to a relatively small value.

Although the probability of cross-shard transaction rollback increases with w, the multi-round verification scheme is only when U = 2, w = 16, which is more efficient than the traditional single-round verification scheme when w = 2. As a result, the probability of a slice transaction being rolled back is much smaller.

This proves that using the multi-round consensus verification scheme to process cross-shard transactions significantly reduces the likelihood of cross-shard transaction rollback and is also conducive to the realization of larger-scale sharding.

Enter the number of shards w = 4, which is the most significant value in cross-shard transactions according to statistics. Table 12.5 analyzes the difference in the probability of transaction rollback across shards when the number of input shards w = 4 and the number of nodes in the shard M keeps growing, using the multi-round verification scheme and not using the multi-round verification scheme.

It can be seen from Table 12.5 that as the number of nodes in a shard increases, the probability of transaction rollback across shards in the two schemes is significantly reduced. However, the verification scheme using multiple rounds is still better than

TABLE 12.4
Influence of Round Number "U" on Cross-Shard Transaction Rollback Probability Q (Rollback)

U	w = 1	w = 2	w = 4	w = 8	w = 16
1	0.092 000	0.180 00	0.320 00	0.540 00	0.790 00
2	0.008 600	0.017 00	0.034 00	0.066 00	0.130 00
3	0.000 790	0.001 60	0.003 20	0.006 30	0.012 00
4	0.000 072	0.000 15	0.000 29	0.000 58	0.001 10
5	6.77E-05	1.35E-05	2.70E-05	5.38E-05	0.000 10
7	5.79E-08	1.15E-07	2.31E-07	4.61E-07	9.16E-07
8	5.35E-09	1.02E-08	2.14E-08	4.26E-08	8.47E-08

TABLE 12.5
Influence of Round Number "T" on Cross-Shard Transaction Rollback Probability Q (Rollback)

U	M = 60	M = 120	M = 200	M = 300	M = 400
1	3.22E-01	9.72E-02	2.73E-03	3.01E-03	5.32E-04
2	3.40E-02	2.54E-03	1.82E-04	2.24E-06	7.10E-08
3	3.11E-03	6.32E--05	1.25E-06	1.67E-09	9.46E-12
4	2.90E-04	1.59E-06	8.47E-09	1.25E-12	1.26E-15
5	2.70E-05	3.98E-08	5.74E-11	9.34E-16	1.69E-19

TABLE 12.6
Values of the Upper Limit of the Number of Rounds Tmax under Different "M" and "g"

L	g = 0.15	g = 0.25	g = 0.30
60	3	8	16
120	3	6	12
200	1	4	10
400	1	3	8
600	1	3	6

the one with various nodes in a single game when the number of nodes is small. As a result, the probability of cross shard transaction rollback is much lower.

The number of nodes in the shard decreases, the size of the shard increases and the size of the chip has increased to N (N >> k);

For the same rollback probability, the required number of rounds U (U << N) enables the overall TPS to be increased by a factor of N/U. Therefore, the multi-round consensus verification scheme can reduce the probability of transaction rollback, expand the scale of sharding, and improve the throughput of the entire blockchain system.

12.3.2.3 The Upper Limit of the Number of Rounds

To limit the upper limit of the delay and prevent the infinite circular consensus of a transaction from affecting the system's overall performance, the upper limit of the number of rounds is set. The upper limit of the number of games T is the maximum number of rounds of consensus, the number of matches when the cross-shard transaction rollback probability is less than 10–8.

According to Table 12.4 and Figure 12.3, when the number of shard nodes and the Byzantine ratio are fixed, the increase of w has little effect on the number of rounds that make the rollback probability of the transaction less than 10–8. Therefore, Table 12.6 analyzes and calculates when the number of input shards w = 4 (which can

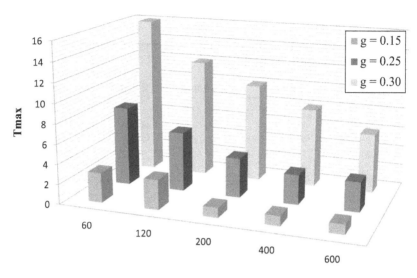

FIGURE 12.3 Upper limit.

roughly represent the number of different input shards w), under the number of nodes M in other bits and the proportion of Byzantine nodes g, the cross shards.

The probability of transaction rollback is less than the upper limit of the number of rounds Umax when 10–8.

When Umax rounds of consensus have continuously verified a transaction, it cannot. If a consensus is reached on the same transaction, the transaction will be abandoned, and the system will be protected as Figure 12.3.

12.3.3 NODE RANDOM ALLOCATION ALGORITHM

When there are too many shard Byzantine nodes and the shard invalidation verification times out, it is necessary to re-allocate a group of nodes to the failed shards through the node random allocation algorithm. This algorithm needs to ensure the high randomness and unpredictability of node allocation.

VRF (verifiable random function, verifiable random function) has the above characteristics, so this scheme selects the VRF function as the random function of the node random allocation algorithm (Williams et al., 2019).

VRF contains 3 functions: VRFG, VRFF, and VRFV.

a) VRFG: public and private essential generation function, which randomly generates a pair of asymmetric encryption keys according to the elliptic curve function, that is, a pair of public and private keys (Q_k, T_k).
b) VRFF: random number and proof generation function, according to the input (T_k and an input x), output two strings, random number value = G1 (T_k,x) and

zero-knowledge proof function proof = G2 (Tk,x), where for any input x, the value values generated by different call nodes are deterministic and unique.
c) VRFV: Verification function, the calling node verifies the value of the random number based on the broadcast input x and the zero-knowledge proof, combined with the public key Qk that generates the lucky number.

The nodes need to be redistributed according to the valid and unique random number value for verification; the value is mapped to a fixed-length output through a hash function. Then, the verification node can enter the corresponding shard according to the result.

The input x in VRF is the seed parameter for generating random numbers, which guarantees the unpredictability of the function. x needs to be public, spontaneous and constantly updated. Accordingly, the VRF input x is set as shown in formula (12.6):

$$x = G\left(C^{g-1}, \{Mk\}\right) \qquad (12.6)$$

Among them, g is the gth block, C^{g-1} is the hash value of the previous block of the current block height, {Mk} is the number obtained by each participating node exchanging a number with the adjacent node in the network a collection of numbers, it can be concluded that x will change every round, with good unpredictability.

12.4 EXPERIMENTAL SETUP AND RESULT ANALYSIS

The existing PBFT consensus cross-sharding schemes all have the phenomenon that the validity of sharding verification impacts the rollback of cross-sharding transactions to verify that the system in this chapter can reduce the probability of cross-shard transaction rollback.

The delay sacrificed by multiple rounds of verification will not reduce the overall transaction throughput TPS of the system. The schemes with better effect among the sharding schemes are selected, respectively.

Cheng et al. (2019) the proposed Omniledger sharding scheme that supports cross-sharding transactions and adopts the PBFT-type consensus algorithm (i.e. ByzCoinX) in the shard is based on the multi-round consensus verification scheme the transaction verification rate and throughput of various projects, and the difference after using multiple rounds of verification.

12.4.1 Experimental Environment and Parameter Settings

In this experiment, the P2P test network built by the laboratory LAN is used as the network environment required for the investigation. Ten shards are constructed through 10 servers in the laboratory. Different port numbers of other servers are used to simulate the creation of varying consensus nodes, and the simulation is performed with ordinary PCs.

For the client that initiates the transaction request in the network, the specific hardware and software environments are as follows.

a) Hardware environment: 8 servers in the LAN: CPU i5 quad-core, RAM 16 GB, Disk 500 GB; another 2 servers: CPU i5 octa-core, RAM 16 GB, Disk 1 TB.
b) Software environment: Operating system: Linux, development environment: Goland. In the calculation in Section 2.2.1, when the number of nodes L in the shard exceeds 60, the probability of collusion attack is lower than 3.2×10^{-8}, so L is set to 60 in the experiment.
- The Omniledger scheme with better cross-shard transaction processing capability is compared with the system using multiple verification rounds.
- The differences in transaction verification rate and throughput are observed.
- The experimental data is drawn with MATLAB. In the design scheme of this chapter, some parameter conditions need to be set before the experiment, as follows:

The attributes of a node are variable. An honest node in this round can become a Byzantine node in the next game.

Set the total number of nodes M = 600, the number of nodes in the shard O = 60, the nodes can dynamically join or exit, and the number of nodes remains unchanged during the sharding.

12.4.2 Experiment Design and Result Analysis

12.4.2.1 Transaction Verification Rate Test

Set the total number of nodes M = 600 and the number of nodes in the shard O = 60 in the blockchain network, verify the transaction verification rate under the two schemes.

Still, the same 50,000 transactions into the two schemes, respectively, set the number of input shards w for each transaction as 1, 2, 4, 6, and 8. Observe the changes in the cross-shard transaction verification rate of the two schemes. As shown in Figure 12.3 with Table 12.7.

According to Figure 12.4, it can be seen that the verification rate of the transaction of the two schemes decreases with the increase of the number of input shards involved in the transaction. Using the multi-round consensus verification scheme, the transaction verification rate is significantly higher than the single-round Omniledger plan.

TABLE 12.7
Transaction Verification Rate

Serial	Multiple Rounds	Omniledger
1	0.99	1
2	0.98	0.93
4	0.95	0.95
6	0.93	0.98
8	0.9	0.99

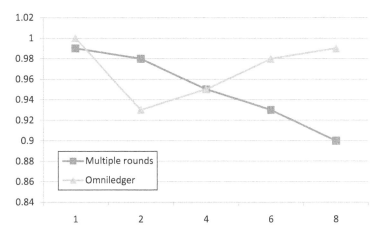

FIGURE 12.4 Transaction verification rate.

This is because the multi-round verification scheme can avoid cross-sharding caused by fragmentation failure with a slight delay. The issue of transaction rollback allows the cross-shard transaction of an invalid shard to continue to be verified in the next round without rolling back the transaction and waiting for resubmission.

In the Omniledger solution, if a single bit fails, intra-shard transactions cannot be verified and processed in a limited time. Therefore, cross-shard transactions need to be rolled back, reducing the verification rate. This experiment can show that the multi-round consensus verification scheme can increase the transaction verification rate, reducing the transaction rollback probability.

12.4.2.2 Transaction Throughput Test

Transaction throughput means the number of transactions processed by the system server per unit of time, generally expressed as TPS (transaction per second, the number of transactions per second), as shown in formula (12.7):

$$TPS = SumTransaction \Delta u / \Delta u \quad (12.7)$$

In formula (12.7), Δu represents the time interval after the transaction is sent to the network until the marketing is on-chain, and $SumTransaction\Delta u$ represents the total number of transactions packaged into the block on-chain during this time interval. Set the total number of nodes $M = 600$ and the number of nodes in the shard $L = 60$ in the blockchain network, verify the transaction throughput under the two schemes, and the system runs 5, 10, 30, 60, 120, respectively 300 min Statistics on transaction processing and calculation of the average TPS for this period.

Observe the changes of the two schemes over time and the evolution of TPS, and calculate the average TPS according to formula (12.7), as shown in Figure 12.5 with Table 12.8.

In Experiment 2, it can be seen that the TPS of the multi-round consensus verification scheme is slightly higher than the original single-round Omniledger scheme.

TABLE 12.8
TPS Performance of the Two Schemes

Serial	Multiple Rounds	Omniledger
1	370	360
2	373	362
4	376	365
6	378	367
8	380	350

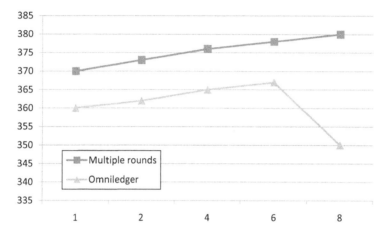

FIGURE 12.5 TPS performance of the two schemes.

This is because Byzantine nodes will cooperate and act together when the system is running, which affects the validity of transaction verification of a single shard.

The method of adding multiple rounds can quickly recover, and the redistributed nodes continue to seek confirmation of transactions in the next game, so the performance of TPS is slightly better.

In the Omniledger scheme of single-round verification transactions, the failed shards cannot be quickly recovered in a short time, and the cross-shard transactions are rolled back, which affects the practical verification of cross-shard transactions, so the average TPS is low.

In the multi-round consensus verification scheme, when the system is running, redistributing nodes to the failed shards will cause a short delay, slightly reducing the transaction throughput as Figure 12.5.

However, the overall performance is still somewhat higher than that of the scheme without multiple rounds, which improves the transaction throughput of the system.

The two experiments comprehensively in Figure 12.5 shows that the multi-round consensus verification scheme that reduces the number of nodes in the shard and increases the scale of the chip is more effective than the single-round verification scheme with more nodes in the bit and a small number of fragments. It reduces the probability of cross-shard transaction rollback, improves the verification rate of transactions, and improves the overall TPS of the system.

12.5 CONCLUSION

For the sharding scheme using the PBFT consensus algorithm, the cross-shard transaction rollback is caused by the uneven distribution of Byzantine nodes after sharding, leading to cross-sharding transactions' rollback.

Among the sharding schemes, the Omniledger scheme with the best comprehensive effect and the widest application range is selected to conduct a simulation comparison experiment. The difference between the two schemes is compared by analyzing the experimental results of transaction verification rate and transaction throughput.

Experiments show that the sharding scheme using multiple rounds of verification can improve the verification rate of transactions and reduce the probability of rollback and resubmission of cross-shard transactions while supporting a larger shard scale.

This ensures data consistency and improves the overall TPS of the sharding system. Sharding technology is the most promising solution to the expansion problem, and the effective processing of cross-shard transactions is an essential prerequisite for ensuring the performance of the sharding system. Therefore, this chapter has a specific reference value for researching many blockchain sharding pros.

REFERENCES

Amiri, M.J., D. Agrawal, & A. El Abbadi, On sharding permissioned blockchains, *2019 IEEE International Conference on Blockchain (Blockchain)*, 2019, pp. 282–5, doi: 10.1109/Blockchain.2019.00044

Bhambri, Khang, A., Sita Rani, & Gaurav Gupta, *Cloud and Fog Computing Platforms for Internet of Things*. Chapman & Hall. ISBN: 978-1-032-101507, 2022, https://doi.org/10.1201/9781032101507.

Bucea-Manea-Țoniş, R., Martins, O., Bucea-Manea-Țoniş, R., Gheorghiță, C., Kuleto, V., Ilić, M. P., & Simion, V. E. *Blockchain Technology Enhances Sustainable Higher Education. Sustainability*, 2021, 13(22), 12347.

Chen, Hong, Z., S. Guo, & P. Li, Pyramid: A layered sharding blockchain system, *IEEE INFOCOM 2021 – IEEE Conference on Computer Communications, 2021*, pp. 1–10, doi: 10.1109/INFOCOM42981.2021.9488747.

Emilie, H. *Investigating the Potential of Blockchains*. 2016-12-11. http://blockchain. Open. ac. uk2016-1-22 (2016).

Gresch, J., Rodrigues, B., Scheid, E., Kanhere, S.S., Stiller, B. The proposal of a blockchain-based architecture for transparent certificate handling. In *Proceedings of the 21st International Conference on Business Information Systems*, BIS 2018; Abramowicz, W., Paschke, A., Eds.; Springer Nature: Basingstoke, UK, 2019, pp. 185–96.

Khang, A., Sita Rani, Meetali Chauhan, & Aman Kataria, IoT equipped intelligent distributed framework for smart healthcare systems, *Networking and Internet Architecture*, 2021, https://arxiv.org/abs/2110.04997v2, https://doi.org/10.48550/arXiv.2110.04997

Khang, A., Geeta Rana, Ravindra Sharma, Alok Kumar Goel, & Ashok Kumar Dubey, The role of artificial intelligence in blockchain applications, *Reinventing Manufacturing and Business Processes Through Artificial Intelligence*, 19–38, 2021, https://doi.org/10.1201/9781003145011

Khang, A., Rani, S., & Sivaraman, A.K. (Eds.). *AI-Centric Smart City Ecosystems: Technologies, Design and Implementation* (1st ed.). CRC Press, 2022. https://doi.org/10.1201/9781003252542

Khang, A., Nazila Ali Ragimova, Vugar Abdullayev Hajimahmud, & Abuzarova Vusala Alyar, Advanced technologies and data management in the smart healthcare system. In *AI-Centric Smart City Ecosystems: Technologies, Design and Implementation* (1st ed.) CRC Press, 2022. doi: 10.1201/9781003252542-16

Kirilova, D., Maslov, N., & Astakhova, T. Prospects for the introduction of blockchain technology into a modern system of education. *International Journal of Open Information Technologies*, 2018, 6(8), 31–7.

Kosba, A., Miller, A., Shi, E., Wen, Z., & Papamanthou, C. Hawk: The blockchain model of cryptography and privacy-preserving smart contracts. *Security & Privacy*, 2016, 839858.

Li, M. & Y. Qin, Scaling the blockchain-based access control framework for IoT via sharding, *ICC 2021 – IEEE International Conference on Communications*, 2021, p. 16, doi: 10.1109/ICC42927.2021.9500403

Liu, Y., J. Liu, Q. Wu, H. Yu, H. Yiming, & Z. Zhou, SSHC: A secure and scalable hybrid consensus protocol for sharding blockchains with a formal security framework. In *IEEE Transactions on Dependable and Secure Computing*, doi: 10.1109/TDSC.2020.3047487

Liu, Y., H. Sun, X. Song, & Z. Chen, OverlapShard: Overlap-based sharding mechanism, *2021 IEEE Symposium on Computers and Communications (ISCC)*, 2021, pp. 1–7, doi: 10.1109/ISCC53001.2021.9631476

Mao, C., & W. Golab, Sharding techniques in the era of blockchain, *2021 40th International Symposium on Reliable Distributed Systems (SRDS)*, 2021, pp. 343344, doi: 10.1109/SRDS53918.2021.00041

Mizrahi, A., & O. Rottenstreich, State sharding with space-aware representations, *2020 IEEE International Conference on Blockchain and Cryptocurrency (ICBC)*, 2020, pp. 1–9, doi: 10.1109/ICBC48266.2020.9169402

Mizrahi, A., & O. Rottenstreich, Blockchain state sharding with space-aware representations, *IEEE Transactions on Network and Service Management* 18(2), 1571–83, June 2021, doi: 10.1109/TNSM.2020.3031355

Nguyen, L.N., T.D. Nguyen, T.N. Dinh, & M.T. Thai, OptChain: Optimal transactions placement for scalable blockchain sharding, *2019 IEEE 39th International Conference on Distributed Computing Systems (ICDCS)*, 2019, pp. 525–35, doi: 10.1109/ICDCS.2019.00059

Ocheja, P et al. Connecting decentralized learning records: a blockchain based learning analytics platform. In *Proceedings of the 8th International Conference on Learning Analytics and Knowledge* (pp. 265–9). ACM, 2018.

Redman J. *MIT Media Lab uses the bitcoin blockchain for digital certificates*, 2016, 12.

Ren, L., & P.A.S. Ward, Understanding the transaction placement problem in blockchain sharding protocols, *2021 IEEE 12th Annual Information Technology, Electronics and Mobile Communication Conference (IEMCON)*, 2021, pp. 0695–0701, doi: 10.1109/IEMCON53756.2021.9623200

San, A.M., Chotikakamthorn, N., & Sathitwiriyawong, C. (2019) Blockchain-based learning credential verification system with recipient privacy control. Presented at the *2019 IEEE International Conference on Engineering, Technology and Education, TALE, Yogyakarta, Indonesia, 10–13 December 2019*.

Serranito, D. et al. (2020) Blockchain ecosystem for verifiable qualifications. Presented at the *2nd Conference on Blockchain Research and Applications for Innovative Networks and Services, BRAINS, Paris, France, 28–30, 2020.*

Sharples, M., & Domingue, J. The blockchain and kudos: A distributed system for educational record, reputation and reward. In *European Conference on Technology Enhanced Learning* (pp. 490–6). Springer, Cham, 2016.

Shobanadevi, A., Tharewal, S., Soni, M., et al. Novel identity management system using smart blockchain technology. *Int J Syst Assur Eng Manag*, 2021. https://doi.org/10.1007/s13 198-021-01494-0

Sohrabi, N., & Z. Tari, ZyConChain: A scalable blockchain for general applications, *IEEE Access.* 8, 158893–910, 2020, doi: 10.1109/ACCESS.2020.3020319

Soni, M., & D.K. Singh, Blockchain implementation for privacy preserving and securing the healthcare data, *2021 10th IEEE International Conference on Communication Systems and Network Technologies (CSNT)*, 2021, pp. 729–34, doi: 10.1109/CSNT51715.2021.9509722

Sun, H. et al. (2018). Application of blockchain technology in online education. *International Journal of Emerging Technologies in Learning*, 13(10), 252–59. doi:10.3991/ijet.v13i10.9455

Tao, Y., B. Li, J. Jiang, H. C.Ng, C. Wang, & B. Li, On sharding open blockchains with smart contracts, *2020 IEEE 36th International Conference on Data Engineering (ICDE)*, 2020, pp. 1357–1368, doi: 10.1109/ICDE48307.2020.00121

Vidal, F., Gouveia, F., & Soares, C. Analysis of blockchain technology for higher education. Presented at the *2019 International Conference on Cyber-Enabled Distributed Computing and Knowledge Discovery, CyberC, Guilin, China,* October 17–19, 2019.

Wang, C., & N. Raviv, Low latency cross-shard transactions in coded blockchain, *2021 IEEE International Symposium on Information Theory (ISIT)*, 2021, pp. 2678–83, doi: 10.1109/ISIT45174.2021.9518047

Wang, G., RepShard: Reputation-based sharding scheme achieves linearly scaling efficiency and security simultaneously, *2020 IEEE International Conference on Blockchain (Blockchain)*, 2020, pp. 237–46, doi: 10.1109/Blockchain50366.2020.00037

Williams, P. Does competency-based education with blockchain signal a new mission for universities? *Journal of Higher Education Policy and Management*, 41(1), 104–17, 2019.

Yang, Z., R. Yang, F.R.Yu, M. Li, Y. Zhang, & Y.Teng, Sharded blockchain for collaborative computing in the Internet of Things: Combined of dynamic clustering and deep reinforcement learning approach, *IEEE Internet of Things Journal*, 2017, doi: 10.1109/JIOT.2022.3152188

Zhang, P., M. Zhou, J. Zhen, & J. Zhang, Enhancing scalability of trusted blockchains through optimal sharding, *2021 IEEE International Conference on Smart Data Services (SMDS)*, 2021, pp. 226–33, doi: 10.1109/SMDS53860.2021.00037

Zheng, P., Q. Xu, Z. Zheng, Z. Zhou, Y. Yan, & H. Zhang, Meepo: Sharded consortium blockchain, *2021 IEEE 37th International Conference on Data Engineering (ICDE)*, 2021, pp. 1847–52, doi: 10.1109/ICDE51399.2021.00165

13 IoT, AI, and Blockchain
An Integrated System Investigation for Agriculture and Healthcare Units

Mandeep Singh, Ruhul Amin Choudhury, and Sweta Chander

CONTENTS

13.1	Introduction		189
13.2	IoT in Agriculture and Healthcare		191
	13.2.1	Agriculture	192
		13.2.1.1 Water Management and Monitoring	193
		13.2.1.2 Soil Monitoring and Assessment	194
		13.2.1.3 Crop Growth Assessment	194
		13.2.1.4 Disease Monitoring	195
		13.2.1.5 Environmental Condition Monitoring	195
	13.2.2	Healthcare	195
		13.2.2.1 Data Acquisition	196
		13.2.2.2 Data Processing	196
		13.2.2.3 Application Phase	196
13.3	AI in Agriculture and Healthcare		197
	13.3.1	Agriculture	197
		13.3.1.1 Soil and Irrigation Management	198
		13.3.1.2 Seed Emerging Assessment	198
	13.3.2	Healthcare	199
		13.3.2.1 Disease Prediction and Diagnosis	199
		13.3.2.2 Artificial Intelligence as Life Assistance	200
13.4	Blockchain		200
	13.4.1	Blockchain with AI and IoT in Agriculture	200
	13.4.2	Blockchain with AI and IoT in Healthcare	201
13.5	Conclusion		202
References			202

13.1 INTRODUCTION

The recent trends and rise of technology have shown a great shift toward artificial intelligence, IoT and blockchain and have perhaps experienced an unbelievable output with the deployment. With rapid evolution in communication devices, IoT, data analytics, artificial intelligence, blockchain has shown great influence in every sector and thus provided a new vision and aspect for every field (Bhambri et al., Cloud & IoT, 2022).

Though they have created buzz in each sector, naturally agriculture and the health sector would be more areas of interest for researchers by maximizing the utilization of these information networks in sustainably feeding the planet and overall health maintenance of the population.

IoT is a system where real world entities are connected to each other possessing an unique id by each connected device and thus allowing communication between them (Ashton et al., 2018).

Increase in communication between devices led to the increase in data, thus the concept of blockchain comes into play. Blockchain is termed as the decentralized technology that keep records of the generated data and its origin. Artificial intelligence technology allows a machine to act and respond like a human (Kaplan et al., 2019).

Summing up of these three technologies and deploying them in various sectors such as agriculture, supply chain, health care, food, retail etc. allows the reforming and development of that industry to the next level.

The Internet of Things allows a wide range of communication between the physical and digital world thus popularly IoT is termed internet of everything. The deployment of IoT in industry is popularly known as Industrial Internet of Things (IIoT). The IoT foundation block comprises data acquisition unit and the communication device.

The acquisition unit collects information from the surrounding environment and communicates it to the processing /control unit (Lee et al., 2015). The IoT platform acts as a bridge between the sensing unit and the data network, where each connected device is continuously communicating with each other. IoT offers a wide range of applications such as smart cities, health care, supply chain, industrial control and agriculture (Khang et al., IoT & Healthcare, 2021).

Talking about the agriculture, IoT has influenced growth in agriculture to a great extent including smart farming (Masner et al., 2016). The integration of IoT has overcome many crucial challenges by inspecting various complications and thus opening the door to green revolution in agriculture (Ray et al., 2017). Also in recent years IoT has shown a great influence in the healthcare unit: generally IoT is deployed to monitor the activity and health orientation of the patient (Kodali et al., 2015). The only IoT based agriculture system or health care unit ends up in monitoring and attaining data about the same. Such data may be directly visible to third parties resulting in a tendency toward privacy breaching (Bergmann et al., 2016). So privacy and security concern is one of the major challenges faced by IoT (Yin et al., 2019).

The concept of blockchain provides security toward the gathered data and transparency in transaction of data. Blockchain is a series of blocks where each block possesses information. Each block stores the data and after the storing of data a new block is generated for the next set of information.

Blockchain technology is timestamp based data which enables security of the database and there is no way to change the timestamp of the digital data, thus making the system secure and reliable and also allowing the sharing or transfer of data from peer-to-peer with a great transparency by initiating a direct transfer of data between two parties without any interference or involvement of third parties allowing peer-to-peer transaction (Ali et al., 2019). So as we know sensing is the origin of data in IoT which security and transparency is managed by blockchain, but in between the data acquisition and peer-to-peer transaction, there is a requirement to extract important features from the data through programming model or statistical approaches or else the huge amount of data acquired at sensory level would have make everything very complex and this feature extraction task is performed by artificial intelligence.

The AI framework requires designing and development of theory through which the machine can exhibit similar intelligence to that of a human being such as perceiving information from the surrounding environment and decision making.

In Khang et al., AI-Centric Smart City Ecosystem (2022) AI is defined as "a system's ability to interpret external data correctly, to learn from such data, and to use those learnings to achieve specific goals and tasks through flexible adaptation." Thus AI manipulates data received from IoT and other perception level and identifies patterns by deploying set based rules or machine learning to achieve a set of objectives as decision making outcome of the acquired data.

So a system can be said to be truly intelligent if it is able to acquire knowledge from the perceived data, and set objectives and priorities defining a set of rules and making decisions with less risk.

In this proposed chapter an investigation is carried out on how the paradigm of agriculture and healthcare has changed due to the incorporation of IoT, AI and blockchain.

13.2 IOT IN AGRICULTURE AND HEALTHCARE

Investigation on influence of IoT in agriculture and healthcare.

The Internet of Things is a technology that allows connectivity between objects by connecting each individual object with a single network enabling successful communication between the objects. The primary objective of the technology is to develop an application for devices, enabling the feature to control and monitoring of a specific domain. IoT has found its wide application starting from industrial processing to smart cities and homes.

IoT connectivity encompasses people, machines, tools and locations, aiming to achieve different intelligent functions from data sharing and information exchange (Bibri et al., 2018). Taking about Agriculture, IoT is deployed to gather information in both the sector including: (a) control; (b) monitor; (c) assessment; (d) management; (e) tracking; (f) operations in supply chain management.

Similarly in the health care sector with deployment IoT sanitary systems can be revolutionized. Also IoT can play a major role in monitoring, assessment of patients, hospital management, high quality health care and the newly grown supply chain management (Ray et al., 2019).

The benefits of IoT in agriculture are: ease of input, sustainability, cost efficiency, profitability, food safety. The representation of architectural structure of IoT in agriculture and healthcare are shown in Figures 13.1(a) and 13.1(b) respectively.

13.2.1 Agriculture

According to the text, IoT plays an important role in booming the agricultural sector. The evolved technologies such as vision unit, RFID, LASER scanner, electromagnetic sensor etc. can be deployed to make great innovations in agriculture.

Basically in agriculture IoT exhibits its impact in data communication, precision irrigation, disease detection and many other as shown in Figure 13.1.

The incorporation of IoT in agriculture provides an enormous number of solution comprises of collective monitoring, controlling of the system and tracking which is further explained with several types of monitoring and control such as temperature monitoring, soil health monitoring, humidity, irrigation control, pest control, location, weather monitoring, livestock management etc. as shown in Figure 13.2.

As per the survey it is accumulated that most of the researches and study related to IoT in agriculture is approx. 70% on monitoring, 25% on controlling and 5% on tracking which is shown in pictorial representation in Figure 13.3.

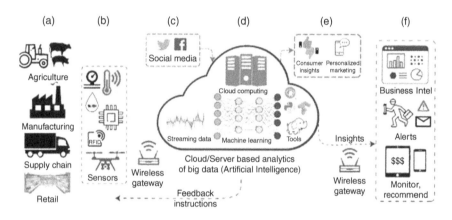

FIGURE 13.1 Architectural framework of IoT.

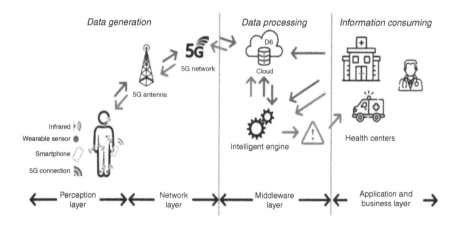

FIGURE 13.2 IoT in healthcare.

IoT, AI, and Blockchain   193

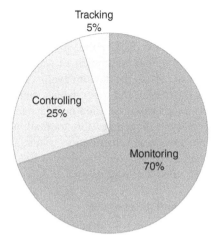

FIGURE 13.3 Pictorial representation.

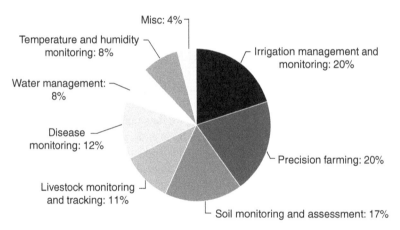

FIGURE 13.4 Pictorial representation of IoT application in agriculture.

As discussed above, the three main applications where IoT is performing a vital role to transform the overall agricultural system. The main classification of the aforementioned section are as follows: (a) irrigation control and management; (b) soil monitoring and assessment; (c) livestock monitoring and tracking; (d) disease monitoringl (e) water management; (f) temperature and humidity monitoring; (g) precision farming and many others as shown in Figure 13.4.

13.2.1.1 Water Management and Monitoring

One of the important sub-domains of agriculture is where incorporation of IoT intends to monitor the requirement of water in the soil, water quality, presence of strong contaminants in water which can change the nature of the soil, pH level of water.

In Postolache et al. (2014), the author proposed a novel IoT based solution to monitor the water quality by taking the measurement of temperature, conductivity and turbidity. The proposed solution is disposed with WSN technology and sensing devices to acquire the data of various parameters of water in urban areas. In another work presented in (Xijun et al., 2014) the author developed a WSN based monitoring device which will monitor the rainfall and water management in the irrigation system.

In Fourati et al. (2014) a web based decision making support system is presented which utilizes sensors to monitor the temperature, rainfall, water level and irrigation system in olive fields. The newly water distribution system allows the irrigation concern to alter water distribution to different sections of field as per the requirement predicted by the system based on data acquisition.

13.2.1.2 Soil Monitoring and Assessment

This domain refers to a crucial aspect for a healthy agriculture due the fact that it holds all the nutrients required for a plant to grow up which are then passed on to humans and other animals as a part of the food cycle.

The healthy soils yields healthy and better production of crops but the quality or the health of the soils is greatly influenced by climatic condition, over use of fertilization, presence of nutrients in the field which tend to deteriorate over time and the farmers required to change their fields for the better yield. So it is of utmost importance to regularly monitor the agriculture fields for maintaining adequate level of the said factors for the healthy soil.

The incorporation of IoT allow us to monitor the chemical status, moisture and other nutrients availability in the soil with deployment of highly precised sensors. These readings will be further communicated for data management and analysis layer for analyzing and decision support (Aleksandrova et al., 2019).

In Chen et al. (2014) the author proposed a soil moisture monitoring and temperature reading system utilizing WSN technology. Both the proposed system are deployed with multiple communication system including ZigBee, GPRS and internet to communicate the acquired data and allow the user to interact with the system via web application.

13.2.1.3 Crop Growth Assessment

In conventional farming it is very difficult to keep a record of the progress of growth rate of the vegetation. The lack of systematically monitoring the growth rate of plants creates major obstacles to increase the crop productivity rate and quality. It is not possible for farmers to regularly monitor the farmlands, so there are various technologies such as remote sensing unit and UAV which can regularly record the crop environment and fetch the data to the user.

In Lee et al. (2012) the author presented a paper where the agricultural land is being analyzed with the help of mobile sensors. The proposed system is deployed to monitor the growth of the grape plants and control the other viticulture activities.

In another work proposed in Feng et al. (2012), an intelligent system can monitor the apple orchards and provide the user with qualitative suggestions based on the acquired data. The sole purpose of the system is to increase the production yield of apples and decrease the management cost along with monitoring of the estimated growth rate of production. The proposed model is a WSN based system equipped with Zigbee and GPRS module for monitoring and communication of data to user.

13.2.1.4 Disease Monitoring

Crop health is an important aspect for the proper production yield as it has a significant impact on the economic status of the country. The monitoring of the crops should be done in a regular manner so that with time the spreading of disease can be avoided.

In traditional method the health monitoring of crop is carried out by human which is time consuming and requires more cost. Thus the deployment of monitoring the plant using remote sensors and communicating the data with IoT is an efficient and cost effective way to avoid the spreading of disease in plants by acting as per the sensed data. The LOFAR-agro Project is the best example for crop or plant monitoring (Langendoen et al., 2006).

The proposed system continuously monitors the potato crop and protects the crop from fungal diseases by analyzing the captured data of crops. The system also monitors the temperature and humidity of the field surrounding it utilizing WSN technology.

13.2.1.5 Environmental Condition Monitoring

An environmental condition monitoring system is proposed in Khandani et al. (2009) that measures the spatial sampling of humidity sensors using WSN. To determine the behavior of 2D correlation, a historical database is being used in the proposed system. Moreover, another environmental conditions monitoring system is proposed in Luan et al. (2015) that integrates forecasting and drought monitoring musing IoT.

13.2.2 HEALTHCARE

Taking the current situation into consideration of the world and the widespread spreading of infectious diseases such as COVID-19 and along with other factors such as long distance, high-cost and the utmost requirement of Quarantine to avoid further spread of the disease, makes it difficult to visit medical centers and even impossible for person such as elderly person, specially abled person (Singh et al., 2020). Therefore the critical situation demands an extensive affordable critical care technology to cover the need of long term caring and remote monitoring of the patient to provide quality life for the patient and to minimize the financial expenses (Mohammed et al., 2020).

The introduction of IoT in this sector has revolutionized the architecture of the healthcare system by analyzing the data and providing a predictive intelligent system by connecting a number of IoT devices which are acquiring various data of patients such as sugar level of the body, pulse of the patient, and temperature of the body along with other important physiological data.

The convergence of IoT with healthcare system has provided the province of smart management of healthcare system, e-monitoring, and Virtual Patient assistance. But with strong advantages, IoT enabled healthcare system is still struggling to achieve its strong hold in the market and that's because of security concerns, self-learning, privacy, lack of proper hardware system and standardization. In general there are three main phases in the workflow of the HIOT which are as follows.

13.2.2.1 Data Acquisition

In this phase the system collects the data from various sources such as sensors, hardware modules, medical devices and even patients can enter the data through interface. The phase collectively comes under the perception layer and the collected data are then communicated to the Data Processing unit via the Network Layer.

13.2.2.2 Data Processing

This phase consists of the processing of the generated data. The phase utilizing various algorithms such as machine learning, Neural Network will analyze the data and provide an output based on the acquired data.

13.2.2.3 Application Phase

This is the final phase of the workflow. In this layer the output of the analyzed data will be conveyed as a command to a device or via application provide the generated information out of the processed data. The data acts as decision making for the medical teams or as an actuator to many other devices.

As per the recent survey around the inclusion of IoT in Healthcare system it is found that in Farahani et al. (2018) in his work has presented the evolution of the healthcare system from clinic based to IoT healthcare.

Furthermore the author also presented an IoT based architecture comprising cloud computing, hardware devices and network layer to transform the traditional healthcare unit to the intelligent healthcare system.

In another objective the author describes the various IoT based application such as remote health monitoring, anomaly detection, early warning detection along with two case studies. Also the author enlisted the various challenges and barriers of this field which have discussed in earlier section such as data management, scalability, security, privacy, standardization.

Also in Darwish et al. (2017) the author presented a comprehensive review which includes a history of IoT to the details of cloud computing and its application to the healthcare system. After that the author also discussed the various obstacles the healthcare system is facing for the full-fledged deployment of the IoT such as privacy, security, hardware design and standardization.

Dhanvijay and Patil (Dhanvijay et al., 2019) presented a survey that reviewed the latest development and deployment of technologies and their influence in IoT enabled healthcare system. The presented chapter particularly focuses on WBAN and its security aspects. Further Sita Rani et al.,IoT& Healthcare (2021) in his proposed work have provided a brief detailed view of the existing technology of IoT in the healthcare unit considering the field of sensing communication and analytics. The author also addressed the recent trends and application of IoT in health sector along with the challenges faced by the implemented system.

Verma and Sood (Verma et al., 2018) proposed a disease prediction system consisting of IoT based health monitoring system and cloud computation. The disease which is subjected to prediction is further classified into four sub categories and then comparative analysis of the output is carried out on the grounds of accuracy, sensitivity and response time levels. In another work by Kumar et al. (2018) an IoT based health application is integrated with cloud computing which monitors, diagnoses and predicts serious diseases.

Further the work suggests a classification algorithm which analyzes the procured data and predicts diabetes in patient. At the end the author performed a comparative analysis with the existing algorithm on the basis of performance, security and response time.

Sood and Mahajan (2019) proposed a monitoring system based on IoT and Fog Computing to control hypertension diseases. Further with the involvement of artificial neural network the system enables itself to predict the risk level of hypertension. At the end the proposed model is compared with the existing cloud computing based system in terms of accuracy and response time.

In another work Suresh et al. (2019) proposed a novel architecture for HIoT to monitor the progression rate of cancer. With the integration of neural network and decision tree the system shows high accuracy growth prediction rate in comparison to other existing algorithms. Tan and Halim (2019) proposed an IoT enabled health care monitoring system which allows measuring heart rate, blood pressure monitoring and body temperature.

Additionally, the proposed model enables the system to predict diabetes and kidney disease in elderly people by deploying an artificial neural network which results in high accuracy rate of prediction in comparison to conventional systems.

13.3 AI IN AGRICULTURE AND HEALTHCARE

Investigation on influence of AI in agriculture and healthcare.

13.3.1 AGRICULTURE

The recent trends of involvement of artificial intelligence in the agriculture sector is quite evident. The agriculture sector is faced with great challenges such as soil testing, improper soil treatment, diseases and pest infestation, big data, and most importantly the knowledge gap between conventional farming and the technology driven farming when it comes to the increase in the yield production.

The introduction of AI in agriculture is greatly influence by the Internet of Things which we have discussed in previous sections. Due to its high flexibility, accuracy and decision making support, artificial intelligence becomes a strong technology to bring a drastic change in the field of agriculture.

With provision of analyzing the acquired data of soil such as temperature, soil moisture, nutrients content, AI will be able to provide a predictive insight about the suitability of the soil for a particular crop and which crop to cultivate at a particular soil at which point of time in a given year and thus limiting the use less water, fertilizer and pesticides.

Agriculture is a system more prone toward uncertainty due to its high dependency on soil quality, weather, crops diseases, climate change and few others. So in order to cope with this uncertainty the agriculture sector has recently seen the involvement of AI in the following fields such as (i) soil and irrigation management; (ii) seed emerging assessment; (iii) weed detection; (iv) precision farming.

13.3.1.1 Soil and Irrigation Management

The health of the soil directly influences the production of crops. For a farmer a good knowledge of the soil health will enhance the production of crops and also conserve soil resources. The health of soil depends on the agricultural practices and treatment of soil to improve its quality. In traditional practice the application of compost and manure in the soil improves soil porosity and aggregation (Pagliai et al., 2004).

Several soil borne diseases affect the growth and production of crops and vegetables which requires proper assessment and control through soil management (Abawi et al., 2000)

The improper irrigation system also leads to crop failure and quality degradation. The new water distribution system allows the irrigation concern to alter water distribution to different section of field as per the requirement predicted by the system based on data acquisition.

Though the mentioned systems are available, the sensing system – which can regularly and rapidly sense the growth of plant, data analytic system that can provide the decision making model for irrigation scheduling – needs to be matched with the spatial and temporal data requirements to make optimal use of these systems.

Khang et al., Sensors and Tools for Smart City Ecosystem (2022) proposed a rule based expert system to monitor and evaluate the performance of the micro irrigation system. Sicat et al. (2005) in their work presenting a fuzzy logic based system utilizing the knowledge of the farmers, the system is able to predict which land is suitable for a particular crop for a particular time of the year. The analysis and prediction is also based on mapping via a fuzzy system.

13.3.1.2 Seed Emerging Assessment

Seedling emergence monitoring is a very crucial application in agriculture. It actually collects imagery data and analyzes the delay in crop germination or an unsuccessful portion of the field due to environmental effects.

The monitoring and mapping are carried out at an early stage of germination with a very high resolution to identify the portion of the field where germination does not take place successfully (Sankaran et al., 2015). The mapping is generally carried out through RGB imagery and multispectral imagery, where the latter one provides great flexibility in calculating vegetation indices.

Weed detection: Weeds are the undesirable plants that grow in agricultural fields and causing severe problems to the agricultural crops. These unwanted plants in the crop field compete with available resources such as water or even space causing decline in production yield and growth.

To control these unwanted growth of weed in the field, application of herbicides in the whole field irrespective of the areas which are affected by weed or not, remains the prominent option for the farmers. However the excessive spraying of herbicides results in the growth of herbicides resistant weeds which pose threats to the growth of crops and also the excessive herbicides H poses a heavy pollution threat to the environment.

Weed mapping is the most common application in precision farming to counter the aforementioned problem of weed growth (Rasmussen et al., 2013) and drones offer a great advantage in this application due to its high degree of flexibility in spatial resolution.

In this context, the agricultural land is differentiated in management zones that each one receives a customized management, as spraying of herbicides will be carried out in few hot spots. To achieve this goal, it is mandatory to generate an accurate mapping of weed for precise spraying of herbicides.

Disease management: For the maximum production of crop, proper control of disease is a significant aspect. In agriculture plant and animal diseases act as a limiting factor which reduces the rate of productivity of the crops. So to yield maximum from agricultural harvesting prior disease detection and precaution is an utmost necessity.

The involvement of AI in disease control and management is the requirement which includes physical, chemical and biological (BEA, 2001). Disease detection is one of the easiest applications due to the fact that the diseases make the plant undergo visible changes in terms of biophysical and biochemical characteristics of plants.

The AI-based application utilized crop imaging information with the help of a data processing algorithm to identify the above said changes in plants. Therefore, the early detection of diseases in plants enable farmers to take necessary steps to control diseases, thus reducing future losses (Kerkech et al., 2018).

13.3.2 HEALTHCARE

The influence of artificial intelligence in the healthcare sector is quite evident from recent trends and development. Likewise, while IoT which is discussed in the previous section contributes as perception and networking layer of a system, artificial intelligence actually acts as a processing layer in the system.

The data acquired from the perception layer needs to be filtered to extract the required valuable data and these data are further processed by the algorithm to attain the desired objective.

The application of AI in healthcare is generally categorized into four sectors namely: disease prediction and diagnosis; living assistance; information processing; and biomedical research.

13.3.2.1 Disease Prediction and Diagnosis

The most important requirement in introducing AI into healthcare is in diagnosing diseases. In recent years a number of interesting breakthroughs have been carried out in this field which make provisions for healthcare professionals to get early indication of various kinds of diseases (Sajda et al., 2006).

One such major class of diagnostic is based on diagnosing vitro utilizing biosensor and chip. For example gene expression is a crucial tool to detect any abnormalities which can be done with an ML based algorithm that will interpret microarray data to detect diseases.

Another such detection is diagnosis of cancer utilizing microarray data analysis (Molla et al., 2004). Also many works have been practically carried out that suggest the AI based system can predict the rate of survival in cancer patient specifically in colon cancer (Shi et al., 2017).

Researchers have also identified some drawbacks in machine learning and proposed methods to limit those drawbacks (Foster et al., 2014). Another classification of diagnosis is based on AI based medical imaging and signal processing. Such techniques are involved in feature extraction from signals such as electromyography, electrocardiography etc. (Krishnan et al., 2018).

AI has also shown its implementation in portable ultrasonic devices thus making it more advanced and allowing any untrained person to use this powerful tool to diagnose many kind of diseases in development.

13.3.2.2 Artificial Intelligence as Life Assistance

With growing advancement in technology, artificial intelligence can be deployed as assistance to elderly and disabled people with utilization of smart robots and making the way for improving the life quality of people. The neural network algorithm fed with specific identification of facial expression allows the system to activate the command based on expression.

Furthermore the human–machine interface based on facial expression allows the wheelchair to move or act based on expression identification without any type of sensor object attached in the body (Dahmani et al., 2016).

A system named RUDO helps visually impaired people live together with sighted people and work in the fields like informatics and electronics (Rabhi et al., 2018). In another work a system has been proposed to assist a pregnant women with dietary and other advices related to crucial stage of pregnancy (Tumpa et al., 2017).

13.4 BLOCKCHAIN

Both the agriculture and healthcare sectors have recently experienced a steep growth in terms of production, security and outreach with the influence of technology like artificial intelligence and IoT.

Now with the integration of blockchain with the aforementioned technology, the said sectors will create huge opportunities for further growth and business. Blockchain plays a pivotal role in supplying agricultural products and healthcare services from source to destination. The upcoming two sections cover the influence of integration of blockchain with AI and IoT in agriculture and healthcare (Khang et al., IoT & Healthcare, 2021).

13.4.1 BLOCKCHAIN WITH AI AND IoT IN AGRICULTURE

A project titled "Investigation on influence of blockchain with AI and IoT in Agriculture," Blockchain for Agrifood discusses the preliminary necessity of

FIGURE 13.5 Blocks involved in agriculture.

blockchain technology in the agrifood market, specifically the need for supply chain management and its opportunities (Brewster et al., 2017).

In Wageningen Economic Research, Brussels (2019) the author discusses the opportunities of blockchain in the agriculture sector. The author discusses how blockchain allows secure transaction among various untrusted parties. The parties are associated with supply chain management and comprises the raw material producers to the buyers and the parties associated with supermarket shelf.

Blockchain also allows the provision of delivery of goods from the producer to the consumer or from seller to the customer, thus increasing the logistic facility and supplying the locally produced goods and also provides information about its location and certification (Kamilaris et al., 2019). The author also suggested chain of blocks associated in agriculture which is shown in Figure 13.5.

Lin et al. (2018) presented a food traceability system involving blockchain and IoT technology and comprising all parties of the smart agriculture ecosystem. IoT devices work in the data acquisition sector reducing human intervention and providing a good output.

Blockchain offers a scale of information to farmers starting from the current transaction, stock price of goods to the buyer's information. In another work the author suggested a new framework for smart agriculture based on IoT and blockchain. Also a food monitoring system is proposed based on food safety utilizing IoT and blockchain (Devi et al., 2019).

The blockchain generally creates a digital identity of a physical product, thus making it a traceable one which allows it to trace the product from the agricultural field to the food table.

AI offers processing of data acquired from several points in SCM with the help of IoT and provides necessary information and guidance to the farmers on pest control, soil health, harvesting etc.

MYbDAIRY (2019) in their documentation discussed the dairy product influenced by artificial intelligence and blockchain. The documentation briefed how the farm is characterized by efficient production, low price, innovative marketing strategy and distribution system (Singh et al., 2020).

13.4.2 BLOCKCHAIN WITH AI AND IOT IN HEALTHCARE

In a project titled "Investigation on influence of blockchain with AI and IoT in Healthcare," with the implementation of blockchain in healthcare, the traditional healthcare system is experiencing restructuring on the grounds of consistency, effective diagnosis, treatment through safe and secure data sharing. Artificial intelligence and machine learning offers disease prediction, diagnosis and critical health monitoring (Khang et al., AI & Blockchain, 2021).

The blockchain can be implemented in various applications of healthcare such as disease diagnosis, tele health monitoring, e-monitoring, medicine suggestion and personal health care application, allowing the user to have access to their secured data and to carry out transparent transactions.

The newly added application of blockchain in healthcare (Khang et al., AI-Centric Smart City Ecosystem, 2022) which garnered much attention is supply chain management. The healthcare industry will show a huge leap in advancement with the involvement of blockchain, AI and IoT.

13.5 CONCLUSION

The above indicates that the traditional framework of agriculture and the healthcare system is evolving and changing due to the fact that the new evolved technologies have made the framework to undergo various changes ensuring security, efficiency and high productivity.

The Internet of Things makes it easier to gather information and communicate this information to the data center for further processing. The next step which is data processing is carried out by various algorithms of artificial intelligence enabling the system to predict, monitor and command the hardware peripherals as per the decision updated through processing of data.

In both IoT and AI, one thing of major concern for the user is data security and privacy. These concerns are eliminated by the involvement of blockchain which provides the user with data transparency and security in transaction. The integration of all this technology will both take the sector by storm and reduce the present shortcomings in the system.

REFERENCES

Abawi, G. S., & T. L.Widmer. (2000) Impact of soil health management practices on soilborne pathogens, nematodes and root diseases of vegetable crops, *Applied Soil Ecology* 15(1), 37–47.

Ahmed, M. et al. (2019) Big data analytics for intelligent Internet of Things. In *Artificial Intelligence in IoT*: Springer, pp. 107–27.

Aleksandrova, M. (2019) Technologies and IoT have the potential to transform agriculture in many aspects. Namely, there are 5 ways IoT can improve agriculture. *Eastern Peak*.

Ali, A., Latif, S., Qadir, J., Kanhere, S., Singh, J., & Crowcroft, J. (2019) Blockchain and the future of the internet: A comprehensive review. arXiv preprint arXiv:1904.00733.

Ashton, K. (2018) How the term "Internet of Things" was invented, A.D. Rayome, Ed., ed: *Tech Republic*.

BEA, Value added by industry as a percentage of gross domestic product. Available at: https://apps.bea.gov/iTable/iTable.cfm?ReqID=51&step=1#reqid=51&step=51&isuri=1&5114= a&5102=5, 201.

Bhambri, Khang, A., Sita Rani, & Gaurav Gupta (2022) *Cloud and Fog Computing Platforms for Internet of Things*. Chapman & Hall. ISBN: 978-1-032-101507, https://doi.org/10.1201/9781032101507.

Bibri, S.E. (2018) The IoT for smart sustainable cities of the future: An analytical framework for sensor-based big data applications for environmental sustainability, *Sustainable Cities and Society* 38, 230–53.

Chen et al. (2014) Design of monitoring system for multilayer soil temperature and moisture based on WSN. In *Proceedings of the 2014 International Conference on Wireless Communication and Sensor Network, Wuhan, China, 13–14 December*, pp. 425–30.

Dahmani, K, Tahiri, A, Habert, O, & Elmeftouhi, Y. (2016) An intelligent model of home support for people with loss of autonomy: a novel approach. In: *Proceedings of 2016 International Conference on Control, Decision and Information Technologies; 2016 Apr 6–8; St. Julian's, Malta*; pp. 182–5.

Darwish, A., Hassanien, A.E., Elhoseny, M., Sangaiah, A.K., & Muhammad, K., (2017) The impact of the hybrid platform of Internet of Things and cloud computing on healthcare systems: opportunities, challenges, and open problems. *Journal of Ambient Intelligence and Humanized Computing* 10(10), 4151–66.

Devi, M.S., Suguna, R., Joshi, A.S., & Bagate, R.A. (2019, February) Design of IoT blockchain based smart agriculture for enlightening safety and security. In *International Conference on Emerging Technologies in Computer Engineering* (pp. 7–19). Springer.

Dhanvijay et al. (2019) Internet of Things: A survey of enabling technologies in healthcare and its applications. *Comput. Network.* 153, 113–31.

Farahani, B., Firouzi, F., Chang, V., Badaroglu, M., Constant, N., & Mankodiya, K. (2018) Towards fog-driven IoT eHealth: promises and challenges of IoT in medicine and healthcare. *Future Generat. Comput. Syst.* 78, 659–76.

Feng, C., Wu, H.R., Zhu, H.J., & Sun, X. (2012) The design and realization of apple orchard intelligent monitoring system based on Internet of Things technology. In *Advanced Materials Research; Trans Tech Publications: Stafa-Zurich, Switzerland*, Volume 546, pp. 898–902.

Foster, K.R. et al. Machine learning, medical diagnosis, and biomedical engineering research–commentary. *BioMed Eng Online* 13:94, 2014.

Fourati, M.A., Chebbi, W., & Kamoun, A. (2014) Development of a web-based weather station for irrigation scheduling. In *Proceedings of the 2014 Third IEEE International Colloquium in Information Science and Technology (CIST), Tetouan, Morocco, 20–22 October*, pp. 37–42.

Ge, L., Brewster, C., Spek, J., Smeenk, A., Top, J., van Diepen, F., & de Wildt, M.D.R. (2017) Blockchain for agriculture and food: Findings from the pilot study (No. 2017-112). Wageningen Economic Research.

Kamilaris, A., Fonts, A., & Prenafeta-Boldú, F.X. The rise of blockchain technology in agriculture and food supply chains. *Trends in Food Science & Technology* 2019, 91, 640–52. doi:10.1016/j.tifs.2019.07.034

Kaplan, A. & M. Haenlein (2019) Siri, Siri, in my hand: Who's the fairest in the land? On the interpretations, illustrations, and implications of artificial intelligence, *Business Horizons* 62(1), 15–25.

Kerkech, M., Hafiane, A., & Canals, R. (2018) Deep leaning approach with colorimetric spaces and vegetation indices for vine diseases detection in UAV images. *Comput. Electron. Agric.* 155, 237–43.

Khandani, S.K., & Kalantari, M. (2009) Using field data to design a sensor network. In *Proceedings of the 2009 43rd Annual Conference on Information Sciences and Systems, Baltimore, MD, USA, 18–20 March 2009*, pp. 219–23.

Khang, A., Geeta Rana, Ravindra Sharma, Alok Kumar Goel, & Ashok Kumar Dubey. (2021) The role of artificial intelligence in blockchain applications, *Reinventing Manufacturing* and *Business Processes Through Artificial Intelligence*, 19–38, https://doi.org/10.1201/9781003145011

Khang, A., Rani, S., & Sivaraman, A.K. (Eds.). (2022) *AI-Centric Smart City Ecosystems: Technologies, Design and Implementation* (1st ed.). CRC Press. https://doi.org/10.1201/9781003252542

Kodali, Ravi Kishore, Govinda Swamy, & Boppana Lakshmi. (2015) An implementation of IoT for healthcare. In *2015 IEEE Recent Advances in Intelligent Computational Systems (RAICS)*, pp. 411–16. IEEE.

Krishnan, S., & Athavale, Y. (2018) Trends in biomedical signal feature extraction. *Biomed Signal Process Control* 43, 41–63.

Kumar, P.M et al. 2018. Cloud and IoT based disease prediction and diagnosis system for healthcare using Fuzzy neural classifier. *Future Generat. Comput. Syst.* 86, 527–34.

Langendoen, K., Baggio, A., & Visser, O. (2006) Murphy loves potatoes: Experiences from a pilot sensor network deployment in precision agriculture. In *Proceedings of the 20th IEEE International Parallel Distributed Processing Symposium, Rhodes Island, Greece, 25–29 April*.

Lin, H. & Bergmann, N.W. (2016) IoT privacy and security challenges for smart home environments. *Information* 7(3), 44. doi:10.3390/info7030044

Lee, I. & K. Lee (, 2015) The Internet of Things (IoT): Applications, investments, and challenges for enterprises, *Bus. Horizons* 58(4), 431440.

Lee, J., Kang, H., Bang, H., & Kang, S. (2012) Dynamic crop field analysis using mobile sensor node. In *Proceedings of the 2012 International Conference on ICT Convergence (ICTC), Jeju Island, Korea, 15–17 October*, pp. 7–11.

Lin, J. et al. (2018) Blockchain and IoT based food traceability system. *International Journal of Information Technology* 24(1), 1–16.

Luan, Q., Fang, X., Ye, C., & Liu, Y. (2015) An integrated service system for agricultural drought monitoring and forecasting and irrigation amount forecasting. In *Proceedings of the 2015 23rd International Conference on Geo informatics, Wuhan, China, 19–21 June*, pp. 1–7.

Masner, J. et al. (2016) Internet of Things (IoT) in agriculture selected aspects, *Agris On-Line Papers Econ. Inform* 8(1), 8388.

Mohammed, M., Syamsudin, H., Al-Zubaidi, S., Aks, R.R., & Yusuf, E. (2020) Novel COVID19 detection and diagnosis system using IOT based smart helmet. *Int. J. Psychosoc. Rehabil.* 24(7), 2296–2303.

Molla, M., Waddell, M., Page, D., & Shavlik, J. (2004) Using machine learning to design and interpret gene-expression microarrays. *AI Mag* 25(1), 23–44.

MYbDAIRY Farms Ltd. (2019) Artificial intelligence and blockchain based dairy products. https://mybdairy.com/document/MYbDAIRY-Whitepaper.pdf

Pagliai, M., N. Vignozzi, & S. Pellegrini (2004) Soil structure and the effect of management practices. *Soil and Tillage Research* 79(2), 131–43.

Pham, T.D., Wells, C., & Crane, D.I. (2006). Analysis of microarray gene expression data. *Curr Bioinform* 1(1), 37–53.

Postolache, O., Pereira, J.D., & Girão, P.S. (2014) Wireless sensor network-based solution for environmental monitoring: Water quality assessment case study. *IET Sci. Meas. Technol* 8, 610–16.

Rabhi, Y., Mrabet, M., & Fnaiech, F. (2018) A facial expression controlled wheelchair for people with disabilities. *Comput Methods Programs Biomed* 165, 89–105.

Rani, Khang, A., Meetali Chauhan, & Aman Kataria, IoT equipped intelligent distributed framework for smart healthcare systems, *Networking and Internet Architecture*, 2021, https://arxiv.org/abs/2110.04997v2, https://doi.org/10.48550/arXiv.2110.04997

Ray, P.P. (2017) Internet of Things for smart agriculture: Technologies, practices and future direction, *J. Ambient Intell. Smart Environ* 9(4), 395420.

Ray, P.P., Dash, D., & De, D. (2019) Internet of Things-based real-time model study on e-healthcare: device, message service and dew computing. *Comput. Network.* 149, 226–39.

Rasmussen, Jesper, J. Nielsen, F. Garcia-Ruiz, S. Christensen, & J.C. Streibig (2013) Potential uses of small unmanned aircraft systems (UAS) in weed research. *Weed Research* 53(4), 242–8.

Sajda, P. (2006). Machine learning for detection and diagnosis of disease. *Annu Rev Biomed Eng* 8, 537–65.

Sankaran et al. (2015) Field-based crop phenotyping: Multispectral aerial imaging for evaluation of winter wheat emergence and spring stand. *Computers and Electronics in Agriculture* 118, 372–9.

Sicat, R.S., Emmanuel John M. Carranza, & Uday Bhaskar Nidumolu (2005) Fuzzy modeling of farmers' knowledge for land suitability classification, *Agricultural Systems* 83(1), 49–75.

Shi, T.W., Kah, W.S., Mohamad, M.S., Moorthy, K., Deris, S., Sjaugi, M.F., et al. (2017) A review of gene selection tools in classifying cancer microarray data. *Curr Bioinform* 12(3), 202–12.

Singh, R.P., Javaid, M., Haleem, A., & Suman, R. (2020) Internet of things (IoT) applications to fight against COVID-19 pandemic. *Diabetes & Metabolic Syndrome: Clin. Res. Rev.* 14(4), 521–4.

Singh, N., Singh, P., Singh, K. K., & Singh, A. (2020) Diagnosing of disease using machine learning. Accepted for *Machine Learning & Internet of Medical Things in Healthcare.* Elsevier Publications.

Sood et al. (2019) IoT-fog based healthcare framework to identify and control hypertension attack. *IEEE Internet of Things Journal* 6(2), 1920–7.

Suresh, A., Udendhran, R., Balamurgan, M., & Varatharajan, R. (2019) A novel Internet of Things framework integrated with real time monitoring for intelligent healthcare environment. *J. Med. Syst.* 43(6), 165–75.

Tan, E. & Halim, Z.A. (2019) Health care monitoring system and analytics based on Internet of Things framework. *IETE J. Res.* 65(5), 653–60.

Tumpa, S.N., Islam, A.B., & Ankon, M.T.M. (2017) Smart care: an intelligent assistant for pregnant mothers. In: *Proceedings of 2017 4th International Conference on Advances in Electrical Engineering; 2017 Sep 28–30; Dhaka, Bangladesh*, pp. 754–9.

Verma, P., Sood, S.K., & Kalra, S. (2018) Cloud-centric IoT based student healthcare monitoring framework. *Journal of Ambient Intelligence and Humanized Computing* 9(5), 1293–1309.

Xijun, Y., Limei, L., & Lizhong, X. (2009) The application of wireless sensor network in the irrigation area automatic system. In *Proceedings of the 2014 International Conference on Networks Security, Wireless Communications and Trusted Computing, Wuhan, China, 25–26 April*, Volume 1, pp. 21–4.

Yin, X.C., Liu, Z.G., Ndibanje, B., Nkenyereye, L., & Riazul Islam, S.M. (2019) An IoT based anonymous function for security and privacy in healthcare sensor networks. *Sensors (Basel)* 19(14), 3146. doi:10.3390/s19143146 PMID:31319562.

14 Security and Privacy Challenges in Blockchain Application

Khalid Albulayhi and Qasem Abu Al-Haija

CONTENTS

14.1	Introduction	207
14.2	Blockchain Architecture	209
	14.2.1 Distributed Ledger	209
	14.2.2 Consensus Mechanism And Mining	211
	14.2.3 Smart Contract Platform	211
14.3	Literature Survey	213
	14.3.1 Blockchain for E-Government	213
	14.3.2 Blockchain for Industry	213
	14.3.3 Blockchain for Cryptocurrency	214
	14.3.4 Blockchain for IoT	214
	14.3.5 Blockchain for Healthcare	215
14.4	Security of Blockchain	216
	14.4.1 Summary of Blockchain Security Policies	216
	14.4.2 Prominent Features of Blockchain Security	217
14.5	Privacy of Blockchain	218
14.6	Security Issues of Blockchain Technology	220
	14.6.1 The 51% Vulnerability Risk	220
	14.6.2 Double Spending and Exploring Sybil	221
14.7	Conclusion	222
References		222

14.1 INTRODUCTION

Nowadays, conventional security techniques alone, such as cryptographic techniques, are not suitable to preserve and safeguard data integrity in this enormous scale of knowledge and intelligent technologies, thus seriously limiting the adoption of various intelligent applications in the future.

Cryptocurrency and blockchain have become a buzzword in both academia and industry. Blockchain has attracted significant academic/industrial attention exceeding the financial industry in research (Zheng et al., 2018).

Blockchain is just a cryptographically verifiable list of data. Being a distributed, incorruptible, and tamper-resistant ledger database, blockchain can address the critical security issues of the IoT ecosystem, particularly on data integrity and reliability (Kshetri et al., 2017). One of the reasons for the enthusiasm for the blockchain is that blockchain databases do not contain any cryptographic guarantees of integrity (Halpin, et al., 2017).

These guarantees are essential for any database operating in an adversarial environment. As is well known in the information security area, all databases operate mostly in an adversarial environment. A blockchain can replace the database architecture (Khang et al., AI & Blockchain, 2021).

The primary advantage of blockchain technology is that the data is decentralized. A distributed public ledger in which all users have the same data generated using the blockchain is an invasion of privacy in many situations. Still, it is necessary for high-value use-cases such as currency (Betts et al., 2016).

In a blockchain network, applications can send and record transactions (events) in a reliable and distributed fashion. The potential applications of blockchain in IoT, as in the example, involve recording some states (i.e., location changes, temperature, and moisture). Blockchain is rapidly growing in popularity and is used extensively for intelligent applications such as smart contracts, distributed storage, and the IoT ecosystem (Dorri, 2016).

Blockchain is a distributed and tamper-resistant ledger built in different locations to provide decentralized data storage services and work as peer-to-peer networks. It consists of blocks sequentially in distributed networks, and cryptography is used to record and secure transactions or transactional matters (Davidson et al., 2016).

Furthermore, security and privacy on blockchain technology is an emerging domain that needs further research. In 2008, the first blockchain was proposed by Satoshi Nakamoto (Nakamoto et al., 2008) when he proposed and implemented bitcoin cryptocurrency to enable the blockchain to increase the spread of bitcoin (Best, 2021).

Bitcoin considers one of the most successful cryptocurrencies, and it achieved great success, with its market cap reaching $1,000 billion in 2021 (Meiklejohn et al., 2013). Bitcoin was not academic research, but it was implemented standardly. Through bitcoin (open-source code), we can transfer money for millions of users without a bank acting as a trusted third party.

Furthermore, the characteristics of the bitcoin network give the possibility that its system is a permissionless network with resistance to censorship. However, bitcoin's technique struggles to secure some degree of anonymity: even though several users believe it implements anonymous payments (Garay et al., 2015), what makes research into blockchain technology (i.e., bitcoin application) even more difficult is that Nakamoto has not addressed the privacy and security features of bitcoin demonstrably. These security and privacy properties only started to be formalized recently Zheng et al., 2017).

The properties of the standard base of blockchain technology are still under contention. Recently, new blockchains with claims to security and privacy have appeared increasingly quickly. Blockchain records data in the form of connected blocks secured and distributed manner. Blockchain is considered a public ledger, and committed transactions are stored in a list of blocks.

Blockchain is constantly growing by adding new blocks at the end of the chain. To user security and ledger consistency, distributed consensus approaches and asymmetric cryptography have been executed consistently to meet the steady expansion.

Blockchain technology generally has key characteristics of decentralization, persistency, anonymity, and audibility. Blockchain can vastly save costs and enhance efficiency and effectiveness (Merkle, 1987).

Transaction in blockchain represents the basic unit of records. When a new transaction is generated, it automatically is broadcast to the entire blockchain network. Validating the signature attached is utilized by nodes receiving the transaction to verify the transaction and then allocate verified transactions into cryptographically secured blocks. Those nodes in the blockchain network are known as block miners.

Firstly, a consensus problem must be solved in a distributed manner to allow a miner to create a block (Khang et al., AI & Blockchain, 2021).

The miners who have solved the consensus issues can broadcast their new blocks via the network. When a new block is received, miners can still resolve the consensus issue of appending the block to their chains maintained locally at the miners after all the transactions included in the block have been verified.

The block has proven to provide the correct answer to the consensus problem. Each new block includes a timestamp and a link to the previous block in the chains by employing cryptography. All miners can synchronize their chains on a systematic basis. To ensure the consistent ledger shared across the distributed network, specific terms should be defined; For instance, bitcoin blockchain keeps the longest chain and discard the shortest, where there is discrepancy among different chains.

14.2 BLOCKCHAIN ARCHITECTURE

There are two nodes in a permissionless blockchain named either end-users or miners. Each node is also able to be a miner or a free user; miners collaboratively maintain the survival of the blockchain on a peer-to-peer (P2P) network system.

There are several blockchain architectures, but we assume that the four-part system architecture: distributed ledger, consensus and mining mechanism, smart contract platform, and application, which shows in Figure 14.1, is the most obvious in the system description.

14.2.1 DISTRIBUTED LEDGER

Distributed ledger is a decentralized database that maintains and allows all network community members to read/write data in it and record it in a standard format. The ledger layer consists of a chain of connected blocks using the hash function. Blocks are organized chronologically to shape a blockchain, and then a new transaction becomes a new block connected to at last block.

For more clarification of the composition of the block, Figure 14.2 clearly shows the nature of its design. A typical block structure involves a block header and a block body. Nonce, which indicates that the block was created correctly.

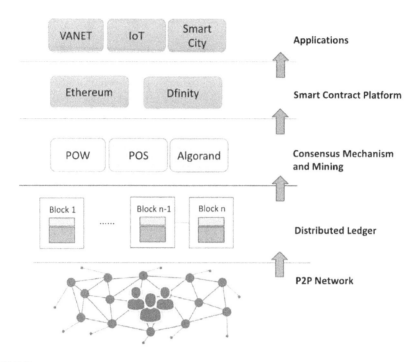

FIGURE 14.1 Blockchain architecture.

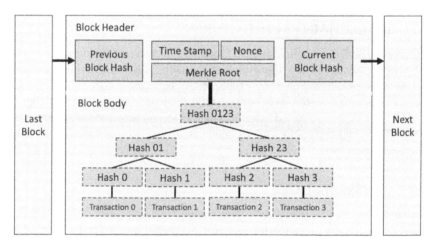

FIGURE 14.2 The structure of the block.

Merkle root tree is responsible for effectively querying and verifying transactions (Bach et al., 2018).

14.2.2 CONSENSUS MECHANISM AND MINING

A consensus mechanism is a fault-tolerant mechanism that uses a set of rules to fulfill essential agreement on a single data value or a single state of the network among distributed nodes (Conti et al., 2018).

The consensus mechanism provides the core functionality to maintain the originality, consistency, and order of the blockchain data transactions across the blockchain network. Generally, a blockchain system and consensus model are secure and a bitcoin application (Xiao et al., 2020).

The security of the consensus layer and blockchain, in general, relies on the assumption of an honest majority; in other words, most of the consensus voting must be honest (Wood et al., 2014).

Some blockchains, such as bitcoin and Ethereum (Solidity, 12-1-2021), are implemented based on an incentive mechanism to motivate miners to create new blocks, which means improving the solidity of the blockchain and enhancing the security of the blockchain.

14.2.3 SMART CONTRACT PLATFORM

Ethereum is considered the first application on which smart contracts are implemented using Solidity language (Christidis et al., 2016). To be easy to understand, smart contracts can be represented as digital upgrading of conventional contracts.

A smart contract is a computer program running on the blockchain network, and it can be executed by itself if and only if the requirements documented in its source code are fulfilled (Xu et al., 2019).

Once the smart contract is written and stored in a blockchain, it cannot be modified or canceled, and trusted by users. This is due to the openness and immutability of the blockchain. This leads to the fact that blockchain can replace many procedures such as human communication supervision costs and using a trusted third party (Zhu et al., 2021).

A smart contract is a computable transaction protocol that can automatically perform contract terms without an intermediary (Peng et al., 2021) and has the characteristics of privacy, enforceability, visibility, and verification. It can achieve self-verification and self-execution after its deployment (Idelberger et al., 2016).

Smart contracts allow users to perform general-purpose calculations on the chain. It consists of four consecutive phases in whole life cycle: (i) creation; (ii) deployment; (iii) execution; and (iv) completion.

They are explained as follows:

1. **Creation**: Several stakeholders first negotiate rights, commitments, and prohibitions on contracts. An agreement can be reached after multiple rounds of discussions and negotiations. Counselors/lawyers assist users (parties) in

drafting an initial contractual agreement. Programmers then transform this agreement, documented in the human languages, into an intelligent contract written in computer languages, including declarative and logic languages (Sillaber et al., 2017). It is like computer software development. It has a life cycle and an iterative process that involves multiple rounds of negotiation and iterationsdEAR, including various parties like stakeholders, lawyers, and programmers.
2. **Deployment**: Smart contracts that are validated can be deployed on blockchains' platforms. More importantly, due to the immutability of the blockchains, contracts stored on them cannot be modified. So, any modification requires the creation of a new agreement. All users (parties) can access the contracts via the blockchains platform once the smart contracts are deployed on blockchains. Furthermore, all parties' digital assets (i.e., smart contract, etc.) are secured by freezing the corresponding digital wallets (Dai et al., 2019).
3. **Execution**: After the deployment has been completed, parties should monitor and evaluate the contractual clauses. The contractual procedures and functions will be automatically executed once the contractual conditions reach all involved parties. It should be noted that a smart contract consists of several declarative statements with logical connections (Koulu et al., 2016). The corresponding declaration will be automatically executed once a condition is triggered; therefore, miners in the blockchain network (Alketbi et al., 2018; Carter et al., 2018).
4. **Completion**: After a contract is executed, the new situations of all concerned parties are updated. During the execution of the contracts, the transactions and updated states are stored in blockchains. In the meantime, digital assets are transferred (for example, money) from one party to another. Therefore, digital assets are unlocked to concerned parties. Consequently, the smart contract has completed the entire life cycle.

Also, the blockchain technologies have been implemented using three major theories: (i) Private-Key Cryptography (ii) decentralized (Peer to Peer) Network (iii) blockchain protocol). The main advantage of blockchain technology is the use of distributed computing methodology, which helps it overcome load sharing problems.

Utilizing distributed computing methodology in building blockchain technology makes blockchain technology highly reliable for storing sensitive information such as transaction processing, medical records, and Secret documentation.

Blockchain technologies consist of Mathematics, Algorithms, Economic and Cryptography models, and distributed consensus algorithms used to solve traditional distributed database synchronization issues. The block body is composed of a transaction counter and transactions.

The maximum number of transactions that a block can contain depends on the block size and the size of each transaction. Furthermore, blockchain utilizes an asymmetric cryptography technique to validate and authenticate transactions (Khang et al., AI-Centric Smart City Ecosystem, 2022).

Digital signature based on asymmetric cryptography is employed in a blockchain network as an untrustworthy environment (Khang et al., AI & Blockchain, 2021).

14.3 LITERATURE SURVEY

Blockchain is widely used to achieve security in several domains, such as EGovernment (Carter et. al, 2018; Al-Jadoori et al., 2019), industry (Al-Jadoori et al., 2019; Al-Haija et al., 2021), cryptocurrency (Monrat et al., 2019; Shi et al., 2020), healthcare (Khan et al., 2020), and the Internet of Things (Alphand et al., 2018). We discuss below some of the approaches in those domains.

14.3.1 BLOCKCHAIN FOR E-GOVERNMENT

Recently, national and authorities have been aggressively discovering innovative technologies to facilitate smooth facilities renovation and to accomplish planned goals such as residents' fulfillment and contentment, services proficiency, and budget optimization.

Blockchain technology has emerged as an attractive technology for several government sectors. The blockchain is a disturbing revolutionary technology performing a crucial position in various areas. Blockchain technology can defeat cybersecurity challenges in cyber-physical systems (CPSs) and their enabled services.

Khang et al., AI &Blockchain (2021) provided an executive study to identify the prospective application cases of blockchain to facilitate government facilities and services. Their assessment demonstrates the enormous potential for blockchain to offer intelligent services and facilities for governments.

Also, Carter et al. (2018) discussed the implications of blockchain technology for citizen trust, privacy, inclusion, and participation. They recommend that governmental administrations require an exhaustive consideration of the blockchain system design ideologies, the conceivable applications in the realm of e-government, and the examination of authority strategies to cope with the restrictions and encounters of the blockchain technology when used in numerous divisions, extending from the economic and commercial field to the communal areas of healthcare and education.

The authors have investigated the influence of blockchain on all levels of government and produced a perception of consequences or applications in society that increase authority concerns.

14.3.2 BLOCKCHAIN FOR INDUSTRY

Al-Jadoori et al. (2019) surveyed different industrial application fields that employs blockchain technology at several levels of the industrial cycle. This investigation for blockchain in industrial application is due to the pragmatic capacities of blockchains in resolving several problems in a variety of industrial fields. This includes securely communicating transactional records, creating effective delivery chain activities, and improving openness.

Blockchain can deal with these concerns through its major features such as distributed, shared, secure, and permissioned transactional ledgers (Al-Jaroodi et al., 2019).

Moreover, authors in Al-Haija et al. (2021) provided a comprehensive survey for blockchain technology in Industries. Their study covered numerous industrial application fields that employ the blockchain to tackle several industrial issues.

They studied the various occasions, advantages, and arguments of integrating blockchain in various industrial services and systems. They also provide a systematic identification model to recognize the requirements that encourage blockchain application for several industrial functions and services.

14.3.3 Blockchain for Cryptocurrency

Al-Haija et al. (2021) propose a classification model to early detect ransomware payments for heterogeneous bitcoin networks. Two supervised machine learning algorithms were utilized to recognize data patterns and construct the classification model, namely Optimizable Decision Trees (ODT) and Shallow Neural Network (SSN).

SSN is a feedforward Multilayer Perceptron (MLP) neural network widely used in pattern recognition and classification tasks in different artificial intelligence and machine learning applications.

Moreover, ODT is commonly used in classification tasks due to its high accuracy. Therefore, SSN is used at the detection stage to flag whether the ransomware exists or not.

Afterward, ODT and SSN classify ransomware attacks into three families, 1) Ransomware–Montreal Family, 2) Ransomware Padua Family, and 3) Ransomware–Princeton Family.

14.3.4 Blockchain for IoT

Researchers proposed several approaches to increase the security in the IoT domain by using blockchain. For instance, ELIB presents an efficient and lightweight integrated blockchain (ELIB) security model for IoT applied to a smart home environment (Alphand et al., 2018).

The model is built on the concept of replacing the resource-intensive Proof of Work (PoW) and Proof of Stake (PoS) mechanisms that are mainly used by cryptocurrencies by a consensus-period approach that restricts the number of new blocks added via the cluster heads (CHs).

The consensus technique has two main goals. The first limited the number of newly created blocks by an arbitrarily selected block generator. Second, increase the arbitrary nature of block creation by enforcing every Overlay Block Manager (OBM) to wait for a random time window before creating the new block.

To avoid the throughput restriction caused by the traditional consensus algorithms, ELIB introduces a Distributed Throughput Management (DTM) to replace the throughput technique used in a typical BC setup. The main goal of DTM is to handle better throughput while verifying that α stays in a particular range ($\alpha_{min}, \alpha_{max}$).

Finally, ELIB utilizes Certificateless Cryptography (CC) as IDbased cryptography to authenticate IoT devices effectively.

Alternatively, IoTChain (Vučinić et al., 2015) proposes a scheme for secure and authorized access to IoT resources that combines Object Security Architecture for the Internet of Things (OSCAR) (Seitz et al., 2017) and Authentication and Authorization for Constrained Environments (ACE) (Esposito et al., 2018).

The basic idea is to make the authorization phase of ACE more flexible and reliable by replacing the authorization server with an authorization blockchain. To have a successful attack on the consensus protocol of the blockchain, the attacker needs to control at least 51% of the blockchain before gaining access to tokens.

The main improvement of IoTChain over ACE is that in IoTChain, the access tokens of the clients are generated based on the access rights described by the resource owner that are stored in smart contracts. However, in ACE, the tickets are transmitted to the client.

Furthermore, IoTChain uses OSCAR (Seitz et al., 2017) to generate personal keys that are transmitted to resource servers over a Datagram Transport Layer Security (DTLS) channel to use them to encrypt and protect the resources. It is worth mentioning that different resources might be encrypted using other keys even if the keys are generated on the same resource server mainly to enforce access control and privileges.

When a client requests access to a certain resource, it sends a request to join the key distribution group associated with the desired help; afterward, the key server checks the contract on the blockchain to verify if the client is authorized or not.

14.3.5 BLOCKCHAIN FOR HEALTHCARE

Recently, blockchain received wide attention in securing several healthcare applications, such as public healthcare management, drug prevention, and clinical trial (McGhin et al., 2019). Researchers think that blockchain can help bridge the gap between privacy and the accessibility of electronic healthcare records (Hussein et al., 2018).

We discuss below some of the proposals in that direction. Ancile introduces a privacy-preserving framework that sends the actual query in a private transaction over HTTPS while creating and storing the hash of the data reference.

Similarly, JP Morgan's Quorum (Kaur et al., 2018) follows a similar approach but lacks elements such as the proxy re-encryption that Ancile uses to streamline the secure transfer of EHRs.

Furthermore, another advantage of proxy re-encryption is that it allows Ancile to store small, encrypted records and keys directly on the blockchain, making it easier to transfer medical records.

One design aspect of Ancile is that data ownership is the patient's right. Therefore, the system does not include any form of mining incentive. Finally, smart contacts manage the access control of the different parties on the blockchain, namely the patients, providers, and third parties.

Esposito et al. proposed that when a new treatment or medical data for a patient comes in, a new block will be created and distributed to all peers in the network (Gupta et al., 2020).

After majority approval of the new block, it will be inserted into the chain. If the support is not achieved, a fork is created, and the block is considered an orphan and not added to the main chain.

Tampering with the data is easily detected because you can't tamper with one block without affecting the subsequent blocks. They also proposed that the medical information be obfuscated or encrypted before placing it into the block.

14.4 SECURITY OF BLOCKCHAIN

Blockchain security is a key concept because it includes protecting data and information used in cryptocurrency transactions and each block against several malicious and harmless attacks. Various policies, tools, and services help detect threats and protect data.

14.4.1 Summary of Blockchain Security Policies

The policies, tools, and services that help to detect threats and protect data in blockchain can be briefly stated as follows:

1. **Defense in penetration**: Many different corrective measures must be enforced to protect data. The principle of protection and defense is that it is supposed to protect data in multiple application layers rather than implement security with a single layer.
2. **Minimum privilege**: Minimum privilege between services and requests should be at the minimum level possible. This approach reduces data access to the lowest possible level to strengthen and improve security.
3. **Manage vulnerabilities**: Vulnerabilities are scanned and managed through modifying authentication of users, and gaps are patched by utilizing of Manage vulnerabilities approach.
4. **Manage risks**: an environment's risks are addressed by first identifying the threat, secondly evaluating the level of risk, and thirdly by monitoring the possibility of continuously occurring hazards.
5. **Manage patches**: the mutable part such as operating system, application, and code should be addressed via testing, installing, and acquiring patches (Lu et al., 2019). Blockchain technology uses various approaches to provide the security required in each block and the transaction. The best one of the techniques used is cryptographic approaches, as it is used in all blockchain applications (for example, in bitcoin). Another philosophy in identifying blockchain security is that the longest chain is legitimate. This dispenses with security opportunities due to 51% control over attack and split issues. Hacked blocks and fake transaction data will eventually die and not make affected.

The blockchain encrypts data utilizing asymmetric encryption methods. Asymmetric encryption has two uses in blockchains: data encryption and digital signatures (Reid et al., 2013).

Data encryption guarantees transaction data security and reduces the risk of transaction data loss or fraud. It is not required to reveal the true identity of the node associated with transaction data. This feature is controversial because it simply helps indirectly with some illegal activities. As much as it is useful in data protection; however, this feature is controversial because it indirectly helps in some unlawful activities (i.e., money laundering) (Karame et al., 2012).

Blockchain technique depends on self-control theory to exchange the data between parties. Blockchain relies on each node to build a strong computation to defend against external attacks without human intervention. Under conditions of complete anonymity, parties can conduct the transaction in an unreliable (untrust) environment with full security (Lu et al., 2018).

After completing the transaction, each node stores the complete data (Zhao et al., 2020). Asymmetric encryption is the primary technology used to secure the blockchain. Asymmetric encryption involves public and private keys (Buldas, Yli-Huumo et al., 2016). Asymmetric encryption uses in two parties in blockchains: data encryption and digital signatures (Aitzhan et al., 2016). Multi-signature technology is a process to solve several challenges security in blockchain where data recorded in the block needs to be verified by other nodes (Wang et al., 2018).

Blind signature technology should be used to fulfill privacy and security requirements. Giving authority to multiple users while securing information is the alternative solution that a decentralized cloud storage system, such as blockchain technology seeks to achieve through smart contracts and cryptographic keys (Eyal et al., 2016). Each node must maintain a backup of the data, which is impossible because of the growing mass data storage (Bhambri et al., Cloud & IoT, 2022).

Although lightweight validation nodes can solve the partly problem, there is still a need to design more effective innovative solutions. Each node stores all historical transaction data, which brings data privacy and performance issues.

Due to the rapid growth in the volume of transactions and data in the blockchain, there is a possibility to solve a problem through the system's scalability. Scalability can be implemented by extending blockchain storage or restructuring the blockchain (Yin et al., 2018).

Especially in transaction authentication, the blockchain technology is based on the elliptic curve digital signature algorithm (ECDSA), which cannot resist the quantum attack in the network (Niranjanamurthy et al., 2019)

For instance, when an attacker obtains a consumer signature or uses the Shor algorithm to derive another user's private key from a public key to sign a set of unauthorized transactions, the legitimate users will lose all their assets and information (Castro et al., 1999).

14.4.2 Prominent Features of Blockchain Security

There are some prominent features of blockchain such that:

- **Decentralization**: The nature of the blockchain depends on the protocol of a peer-to-peer network environment, which means it is suitable for some

conventional Networks and not ideal for others. This feature provides resilient network configurations and reduces risks for single-point failures (SPF).
- **Integrity** and **Immutable** Nodes: Blockchains can keep transactions permanently in a verifiable manner. Precisely, the senders' signatures in transactions guarantee confidentiality, authenticity and integrity, and nonrepudiation on the transactions. Any recorded data cannot be updated, even partly, that is ensured by the hash chain structure of blockchains. Blockchains' protocols can guarantee correct and consistent records in a large percentage. The protocols may allow failures and attacks in limited ranges, for example, attackers with less than 12 hash power in Proof of Work (PoW) (Castro et al., 2002). Any records will be held forever, and it is not changed unless someone has control of more than 51% of nodes simultaneously.
- **Anonymity** and **Secrecy**: Blockchains can use variable public keys as users' identities to preserve and protect anonymity and privacy (Khalilov et al., 2018). This is so important for many applications and services, especially for those that need to maintain anonymity and privacy (Zha et al., 2016). Data transfer (transaction) is anonymous, which means concerned parties are unknown. We only need to know the person's blockchain address.
- **Transparency**: Transparency is available in the data record of each block node, which is a requirement of the blockchain methodology.
- **Open source**: The open-source blockchain systems are available to everyone, which means the record can be reviewed publicly. Furthermore, developers can use blockchain technology to implement their applications
- **Other features** include Cryptographically Secured, Non-repudiable, Auditable, Not Dependencies on Thrustless Third Parties (TTP). Interests in applying blockchain in securing networks have already emerged in academia and industry, intending to provide security (Zyskind et al., 2015).

14.5 PRIVACY OF BLOCKCHAIN

Privacy is the ability of an individual or group to separate themselves or their private information from others during the transmission of data through the transaction process. Security in a blockchain means exchanging and transferring data without losing a reference guide to the data.

Privacy enables users to maintain their connection to the system without explicitly revealing themselves or showing their activity over the entire system. The goal of privacy enhancement in the blockchain is to protect the cryptographic profiles of users or reuse them by different users.

The volume of the differences that occur when applying blockchain technology is very large and cannot be counted, so the security and privacy feature cannot be overlooked.

- **Stored data sorting**: Blockchain provides the ability to store all kinds of data. The opinion of privacy in the blockchain varies between individual and

organizational data. Although privacy rules can be applied to separate data, stringent rules apply only to sensitive and corporate data.
- **Storage distribution**: The nodes that store the complete copy of the blockchain are called full nodes. Redundancy in data usually often is occurred when full nodes combine with the append-only characteristics of the blockchain. It is calculated that this data redundancy in the blockchain techniques creates two features: transparency and variability. Transparency and contrast levels are determined by the level of blockchain application compatibility with data reduction.
- **Append-only**: It is impossible to change the information of the previously undiscovered blocks inside the blockchain. The append-only feature of blockchain cannot be correct for users, especially when the user records incorrect information. Extreme attention should be given while allocating rights to data subjects in blockchain technology.
- **Private versus public blockchain**: From the privacy viewpoint, blockchain considers a great technique. At an advanced level, authorized users can encrypt the restricted data on a block to conditional access since each node in the blockchain holds a copy of the whole blockchain.
- **Non-Permissioned vs. permissioned types of blockchain**: In principle, all users can add data/ information with no constraint through public or unauthorized blockchain applications. Network control can be redistributed by allowing a trusted third party to restore or redistribute it (Lu et al., 2019).

The blockchain needs to meet the following requirements to protect privacy: (i) The connection between blocks (transactions) must not be discoverable or visible, and (ii) Transaction content is known only to concerned parties.

However, everyone can access the blockchain with no restrictions, but two factors must be considered to privacy requirements) (Gramoli et al., 2020):

a) Identity Privacy: it means the inconsistency between transaction scripts and the real identities of its parties.
b) Transaction Privacy:the transaction contents must be accessed only by concerned parties and kept unknown to the blockchain network.

As blockchain works like a public ledger, there needs to ensure several factors, including (a) Protocols for Commitment; (b) Consensus; (c) Security; and (d) Privacy and Authenticity. The various risk and attacks have existed in several blockchain applications, such as privacy leakage, private key security, double spending, balanced attack, and mining attack.

The cryptographic primitives, privacy, and anonymity in the blockchain are key problems at transaction hashing. Diverse cryptographic algorithms have been utilized during the mining process compared to bitcoin and Ethereum algorithms (Alexopoulos et al., 2017).

Authors in Zikratov et al. (2017) examined the worth of using open distributed ledgers (ODLs) to ensure authentication trust Management. Data integrity can be achieved, and threats can be reduced through transactions authentication in a blockchain network (Wang et al., 2020). Authors in Qin et al. (2020) proposed the concept of designated-verifier proof of help to bitcoin transactions using elliptic curve public-key cryptography. Public key infrastructure integrates with the CeCoin PKI scheme to enhance security (Kshetri et al., 2017).

Cybersecurity is another significant matter in the blockchain technique (Militano et al., 2016). The authors in Pinzón et al. (2016) proposed a new trust-based solution to design an efficient cooperative content in cellular environments based on D2D communications.

Some blockchain applications are subject to various attacks, such as Code-based attacks, Dust transactions, Double spend attacks, Double spending (Kshetri et al., 2017), and Online attacks like ransomware (Gervais et al., 2016). Hence, these attacks need to be addressed. Authors in Chanson et al. (2017) proposed PoW blockchain variants and a framework PoW-based deployments to compare the tradeoffs between blockchain security provisions and their performance.

The data generated by the network becomes more important, especially in times of increased capabilities to collect and analyze individual data. In light of these challenges, the authors in Sousa et al. (2018) present that blockchain technology can manage privacy by providing an odometer fraud prevention system.

During the movement of cars, the application stores the data mileage and data GPS and secures it on the blockchain to prohibit odometer fraud. The application infrastructure should be safe against outside accessibility; otherwise, attacks such as ransomware can be possible.

The authors in Natoli et al. (2016) proposed a Byzantine Fault-Tolerant (BFT) ordering service in the Hyperledger Fabric with a new consensus technique to successfully order verified transactions. Blockchain systems (Khang et al., AI & Blockchain, 2021) may fail to ensure consensus without fulfilling the conditions and not transfer repeatable execution on the Ethereum private chain. The authors in identify some situations that may help deal with these challenges.

14.6 SECURITY ISSUES OF BLOCKCHAIN TECHNOLOGY

Although blockchain technologies have provided us with many reliable features and services, security and privacy issues still have a concern that needs attention. Several authors have conducted studies on blockchain technology's security and privacy challenges, but a systematic study is keenly required to cover all the essential factors.

The security challenges are as follows: (i) 51% vulnerability risk; (ii) double spending and exploring sybil; (iii) mining pool attacks; (iv) client-side security threats; (v) forking; (vi) criminal activity; (vii) private key security; and (viii) transaction privacy leakage.

14.6.1 THE 51% VULNERABILITY RISK

Blockchain functions to establish mutual trust between parties by integrating distributed consensus approaches concurrently. The computing power is distributed among the whole network.

Security and Privacy Challenges in Blockchain Application

Data miners' experts check hashes generated by CPU cycles to ensure that hashes are in original condition. When these miners form a single block, the technology will have the most computing power possible.

An attack by a mining pool with 51% or more computing power can compromise blockchain security and compromise the whole cryptocurrency network. When a single miner reaches 50% of the hashing power or more from the total hashing power of a complete blockchain system, the miner easily launches a 51% attack.

Moreover, this attack can modify and exploit the information of the blockchain technique. The vulnerabilities listed below can occur due to this attack.

- Reverse transaction attacks
- Double spending
- Exclude transactions
- Modify transactions
- Disturbing operation of other miners
- Termination of the verification process

14.6.2 Double Spending and Exploring Sybil

For example, when a client (consumer) uses the same cryptocurrency for several transactions in the blockchain application, this is called double-spending. Due to double spending, attackers can perform their attacks called racial attacks.

In blockchain-based on PoW, these attacks are relatively easy to implement because attackers can exploit or confirm the time between the initiations of two transactions.

The attackers can get the outcome of the first transaction, which may lead to double spending before their second transaction gets invalid. A model depiction of an attacker's double-spending behavior (*Assumption/Premise*):

- The agent (seller) address is known to the attackers before the attacker starts attacking.
- Two transactions (T1 and T2) have been starting, we assume that both the transactions have the same bitcoin address as the input the performing technically T1 has been set a recipient address as a non-colluding address for a targeted seller address.
- T2 has been set as a recipient address as a colluding address that an attacker controls.
- T1 will be added to the seller's wallet as a starting point.
- Before T1 gets a confirmation, T2 gets initiated.
- While T2 is in process, an attacker will obtain the confirmation of T1, and the T1 transaction will be completed simultaneously.
- Once T2 is completed, T1 will be mined as invalid.
- When T2 completes, T1 will be mined as invalid when the attacker has already operated the same cryptocurrency twice.
- Due to the colluding address of T2 owned by the attacker, the attacker can still hold the bitcoin (BTC) and has a services application without spending BTC.

14.7 CONCLUSION

Blockchain technology was first proposed in 2008 by Satoshi Nakamoto as a public and distributed ledger that is replicated among several nodes in a peer-to-peer (P2P) network.

The blockchain contains a list of immutable and verified records called "blocks." Each block contains a previous hash to link the previous block because the nature of blockchain is linked to the backlist, also it contains a nonce, transaction root, and network timestamp to indicate the time that a block is added to the chain. The characteristics of the blockchain are immutability and anonymity.

Blockchain security provides solutions in many fields such as cryptocurrency, healthcare, and IoT technology. In this chapter, we have studied the decentralized blockchain technology along with its different features correlating the blockchain system with the diverse security applications and the most recent trends of blockchain security.

Also, several possible security contraventions and susceptibilities of blockchain technology have been investigated and reported. Moreover, an inclusive inspection and evaluation for the e-security services and blockchain technology for improved IoT Security applications and services.

Eventually, blockchain technology is used to provide a distributed database that relies on a P2P network and provides the highest level of trust, availability, and reliability without needing a third trusted party.

REFERENCES

Aitzhan, N.Z., & D. Svetinovic. Security and privacy in decentralized energy trading through multi-signatures, Blockchain and anonymous messaging streams. *IEEE Transactions on Dependable and Secure Computing*, 2016. 15(5), 840–52.

Alexopoulos, N., et al. Beyond the hype: On using blockchains in trust management for authentication. In *2017 IEEE Trustcom/BigDataSE/ICESS*. 2017. IEEE.

Al-Haija, Q.A., & Alsulami, A.A. High performance classification model to identify ransomware payments for heterogeneous bitcoin networks. *Electronics*, 2021. 10(17), 2113.

Al-Jaroodi, J., & N. Mohamed, Blockchain in industries: A survey, *IEEE Access* 7, 36500–15, 2019, DOI: 10.1109/ACCESS.2019.2903554

Al-Jaroodi, J., & N. Mohamed, Industrial applications of blockchain, 2019. *IEEE 9th Annual Computing and Communication Workshop and Conference (CCWC)*, 2019, pp. 0550–5, DOI: 10.1109/CCWC.2019.8666530

Alketbi, A., Q. Nasir, & M.A. Talib, Blockchain for government services – use cases, security benefits, and challenges, 2018, *15th Learning and Technology Conference (L&T)*, 2018, pp. 112–19, DOI: 10.1109/LT.2018.8368494

Alphand, O., et al. (2018, April). IoTChain: A blockchain security architecture for the Internet of Things. In *2018 IEEE Wireless Communications and Networking Conference (WCNC)* (pp. 1–6). IEEE.

Bach, L.M., B. Mihaljevic, & M. Zagar. Comparative analysis of blockchain consensus algorithms. In *The 2018 41st International Convention on Information and Communication Technology, Electronics and Microelectronics (MIPRO)*. 2018. IEEE.

Best, R.d. Daily bitcoin (BTC) market cap history until December 8, 2021. 12-1-2021]; Available from: www.statista.com/statistics/377382/bitcoin-market-capitalization/.
Betts, B. Blockchain technology allows distributed retention of encrypted data and is at the heart of unified cloud storage. 2016 [2021]; Available from: www.computerweekly.com/feature/.
Bhambri, Khang, A., Sita Rani, & Gaurav Gupta, *Cloud and Fog Computing Platforms for Internet of Things.* Chapman & Hall. ISBN: 978-1-032-101507, 2022, https://doi.org/10.1201/9781032101507.
Buldas, A., R. Laanoja, & A. Truu, Keyless signature infrastructure and PKI: hash-tree signatures in a pre-and post-quantum world. *International Journal of Services Technology and Management,* 2016. 23(1–2), 117–30.
Carter, Lemuria, & Jolien Ubacht. Blockchain applications in government. *Proceedings of the 19th Annual International Conference on Digital Government Research: governance in the data age.* 2018.
Castro, M. & B. Liskov. Practical Byzantine fault tolerance. In *OSDI.* 1999.
Castro, M. & B. Liskov, Practical Byzantine fault tolerance and proactive recovery. *ACM Transactions on Computer Systems (TOCS),* 2002. 20(4), 398–461.
Chanson, M., et al. Blockchain as a privacy enabler: An odometer fraud prevention system. In *Proceedings of the 2017 ACM International Joint Conference on Pervasive and Ubiquitous Computing and Proceedings of the 2017 ACM International Symposium on Wearable Computers.* 2017.
Christidis, K. & M. Devetsikiotis, Blockchains and smart contracts for the Internet of Things. *IEEE Access,* 2016. 4, pp. 2292–2303.
Conti, M., et al., A survey on security and privacy issues of bitcoin. *IEEE Communications Surveys & Tutorials,* 2018. 20(4), 3416–52.
Dai, H.-N., Z. Zheng, & Y. Zhang, Blockchain for the Internet of Things: A survey. *IEEE Internet of Things Journal,* 2019. 6(5), 8076–94.
Davidson, S., P. De Filippi, & J. Potts, *Economics of Blockchain.* Available at SSRN 2744751, 2016.
Dorri, A., S.S. Kanhere, & R. Jurdak, *Blockchain in the Internet of Things: Challenges and Solutions.* arXiv preprint arXiv:1608.05187, 2016.
Esposito, C., De Santis, A., Tortora, G., Chang, H., & Choo, K.K.R. Blockchain: A panacea for healthcare cloud-based data security and privacy?. *IEEE Cloud Computing,* 2018.5(1), 31–7.
Eyal, I., et al. Bitcoin-ng: A scalable blockchain protocol. In *13th {USENIX} Symposium on Networked Systems Design and Implementation ({NSDI}* 16. 2016.
Garay, J., A. Kiayias, & N. Leonardos. The bitcoin backbone protocol: Analysis and applications. In *Annual International Conference on the Theory and Applications of Cryptographic Techniques.* 2015. Springer.
Gervais, A., et al. On the security and performance of proof of work blockchains. In *Proceedings of the 2016 ACM SIGSAC Conference on Computer and Communications Security.* 2016.
Gramoli, V., From blockchain consensus back to Byzantine consensus. *Future Generation Computer Systems,* 2020. 107, 760–9.
Gupta, N., A deep dive into security and privacy issues of blockchain technologies. In *Handbook of Research on Blockchain Technology.* 2020, Elsevier, pp. 95–112.
Halpin, H. & M. Piekarska. Introduction to security and privacy on the blockchain. In *2017 IEEE European Symposium on Security and Privacy Workshops (EuroS&PW).* 2017. IEEE.

Hussein, A.F., Arun Kumar, N., Gonzalez, G.R., Abdulhay, E., Tavares, J.M.R., & de Albuquerque, V.H.C. Medical records are managed and secured by a blockchain-based system supported by a genetic algorithm and discrete wavelet transform. *Cognitive Systems Research*, 2018. 52, 1–11.

Idelberger, F., et al. Evaluation of logic-based smart contracts for blockchain systems. *The International Symposium on Rules and Rule Markup Languages for the Semantic Web*. 2016. Springer.

Karame, G.O., E. Androulaki, & S. Capkun. Double-spending fast payments in bitcoin. In *Proceedings of the 2012 ACM Conference on Computer and Communications Security*. 2012.

Kaur, H., Alam, M.A., Jameel, R., Mourya, A.K., & Chang, V. A proposed solution and future direction for Blockchain-based heterogeneous medicare data in a cloud environment. *Journal of Medical Systems*, 2018. 42(8), 1–11.

Khalilov, M.C.K., & A. Levi, A survey on anonymity and privacy in bitcoin-like digital cash systems. *IEEE Communications Surveys & Tutorials*, 2018. 20(3), 2543–85.

Khan, F.A., Asif, M., Ahmad, A., Alharbi, M., & Aljuaid, H. Blockchain technology, improvement suggestions, security challenges on the smart grid, and its application in healthcare for sustainable development. *Sustainable Cities and Society*, 2020. 55, 102018.

Khang, A., Geeta Rana, Ravindra Sharma, Alok Kumar Goel, & Ashok Kumar Dubey, The role of artificial intelligence in blockchain applications, *Reinventing Manufacturing and Business Processes Through Artificial Intelligence*, 19–38, 2021, https://doi.org/10.1201/9781003145011.

Khang, A., Sita Rani, Meetali Chauhan, & Aman Kataria, IoT equipped intelligent distributed framework for smart healthcare systems, *Networking and Internet Architecture*, 2021, https://arxiv.org/abs/2110.04997v2, https://doi.org/10.48550/arXiv.2110.04997.

Khang, A., Vladimir Hahanov, Eugenia Litvinova, Svetlana Chumachenko, Vugar Abdullayev Hajimahmud, & Abuzarova Vusala Alyar, The key assistant of smart city – sensors and tools, *AI-Centric Smart City Ecosystems: Technologies, Design and Implementation* (1st ed.) (2022). CRC Press. https://doi.org/10.1201/9781003252542

Koulu, R., Blockchains and online dispute resolution: smart contracts as an alternative to enforcement. *SCRIPTed*, 2016. 13, 40.

Kshetri, N., Blockchain's roles in strengthening cybersecurity and protecting privacy. *Telecommunications Policy*, 2017. 41(10), 1027–38.

Kshetri, N., Can Blockchain strengthen the Internet of Things? *IT Professional*, 2017. 19(4), 68–72.

Lu, Y., Blockchain and the related issues: A review of current research topics. *Journal of Management Analytics*, 2018. 5(4): 231–55.

Lu, Y., The Blockchain: State-of-the-art and research challenges. *Journal of Industrial Information Integration*, 2019. 15, 80–90.

McGhin, T., Choo, K.R., Liu, C.Z., & He, D., 2019. Blockchain in healthcare applications: research challenges and opportunities. *J. Netw. Comput. Appl.* 135, 62–75.

Meiklejohn, S., et al. A fistful of bitcoins: characterizing payments among men with no names. In *Proceedings of the 2013 Conference on Internet Measurement*. 2013.

Merkle, R.C. A digital signature based on a conventional encryption function. In *Conference on the Theory and Application of Cryptographic Techniques*. 1987. Springer.

Militano, L. et al., Trust-based and social-aware coalition formation game for multihop data uploading in 5G systems. *Computer Networks*, 2016. 111, 141–51.

Monrat, A.A., Schelén, O., & Andersson, K. A survey of blockchain from the perspectives of applications, challenges, and opportunities. *IEEE Access*, 2019. 7, 117134–51.

Nakamoto, S., Bitcoin: A peer-to-peer electronic cash system. *Decentralized Business Review*, 2008: 21260.12-1-2021. Available from: https://bitcoin.org/en/.

Natoli, C. & V. Gramoli. The blockchain anomaly. In *2016 IEEE 15th International Symposium on Network Computing and Applications (NCA)*. 2016. IEEE.

Niranjanamurthy, M., B. Nithya, & S. Jagannatha, Analysis of blockchain technology: pros, cons, and SWOT. *Cluster Computing*, 2019. 22(6), 14743–57.

Peng, L., et al., Privacy preservation in permissionless Blockchain: A survey. *Digital Communications and Networks*, 2021. 7(3), 295–307.

Pinzón, C. & C. Rocha, Double-spend attack models with time advantage for bitcoin. *Electronic Notes in Theoretical Computer Science*, 2016. 329, 79–103.

Qin, B., et al., Cecoin: A decentralized PKI mitigating MitM attacks. *Future Generation Computer Systems*, 2020. 107, 805–15.

Rana, Khang, A., Ravindra Sharma, Alok Kumar Goel, & Ashok Kumar Dubey, *Reinventing Manufacturing and Business Processes Through Artificial Intelligence*, 2021, https://doi.org/10.1201/9781003145011

Reid, F. & M. Harrigan, *An Analysis of Anonymity in the Bitcoin System, in Security and Privacy in Social Networks*. 2013, Springer, po. 197–223.

Seitz, L., Selander, G., Wahlstroem, E., Erdtman, S., & Tschofenig, H. *Authentication and Authorization for Constrained Environments (ACE)*. Internet Engineering Task Force, Internet-Draft draft-IETF-aceoauth-authz-07, 2017.

Shi, S., He, D., Li, L., Kumar, N., Khan, M. K., & Choo, K.K.R. Applications of blockchain in ensuring the security and privacy of electronic health record systems: A survey. *Computers & Security*, 2020, 101966.

Sillaber, C., & B. Waltl, Life cycle of smart contracts in blockchain ecosystems. *Datenschutz und Datensicherheit-DuD*, 2017. 41(8), 497–500.

Sousa, J. et al. Byzantine fault-tolerant ordering service for the hyper ledger fabric blockchain platform. In *2018 48th Annual IEEE/IFIP International Conference on Dependable Systems and Networks (DSN)*. 2018. IEEE.

Vučinić, M., Tourancheau, B., Rousseau, F., Duda, A., Damon, L., & Guizzetti, R. OSCAR: Object security architecture for the Internet of Things. *Ad Hoc Networks*, 2015. 32, 3–16.

Wang, H., D. He, & Y. Ji, Designated-verifier proof of assets for bitcoin exchange using elliptic curve cryptography. *Future Generation Computer Systems*, 2020. 107, 854–62.

Wang, S., Y. Zhang, & Y. Zhang, A blockchain-based framework for data sharing with fine-grained access control in decentralized storage systems. *IEEE Access*, 2018. 6, 38437–50.

Wood, G., Ethereum: A secure decentralized generalized transaction ledger. *Ethereum project yellow paper*, 2014. 151(2014), 1–32.

Xiao, Y., et al. Modeling the impact of network connectivity on consensus security of proof-of-work blockchain. In *IEEE INFOCOM 2020-IEEE Conference on Computer Communications*. 2020. IEEE.

Xu, Q., Z. Su, & Q. Yang, Blockchain-based trustworthy edge caching scheme for the mobile cyber-physical system. *IEEE Internet of Things Journal*, 2019. 7(2), 1098–1110.

Yin, W., et al., An anti-quantum transaction authentication approach in blockchain. *IEEE Access*, 2018. 6, 5393–5401.

Yli-Huumo, J., et al., Where is current research on blockchain technology? – a systematic review. *PloS one*, 2016. 11(10), e0163477.

Zha, X., K. Zheng, & D. Zhang. Anti-pollution source location privacy-preserving scheme in wireless sensor networks. In *2016 13th Annual IEEE International Conference on Sensing, Communication, and Networking (SECON)*. 2016. IEEE.

Zhao, X., et al. Blockchain and distributed systems. In *International Conference on Web Information Systems and Applications*. 2020. Springer.

Zheng, Z., et al., Blockchain challenges, and opportunities: A survey. *International Journal of Web and Grid Services*, 2017. 14(4): 352–75.

Zheng, Z., et al. An overview of blockchain technology: Architecture, consensus, and future trends. In *2017 IEEE International Congress on Big Data (BigData Congress)*. 2017. IEEE.

Zhu, P. et al., Using blockchain technology to enhance the traceability of original achievements. *IEEE Transactions on Engineering Management*, 2021.

Zikratov, I., et al. Ensuring data integrity using blockchain technology. In *2017 20th Conference of Open Innovations Association (FRUCT)*. 2017. IEEE.

Zyskind, G. & O. Nathan. Decentralizing privacy: Using blockchain to protect personal data. In *2015. IEEE Security and Privacy Workshops*. 2015. IEEE.

15 Blockchain-based Cloud Resource Allocation Mechanisms for Privacy Preservation

Akhilesh Kumar, Nihar Ranjan Nayak, Samrat Ray, and Ashish Kumar Tamrakar

CONTENTS

- 15.1 Introduction ... 227
- 15.2 Blockchain Statement ... 230
- 15.3 Blockchain-Based Cloud Resource Allocations ... 231
  - 15.3.1 Manufacturing Resource/Requirement Release ... 232
  - 15.3.2 The Matching Process ... 233
    - 15.3.2.1 Step 1 ... 234
    - 15.3.2.2 Step 2 ... 234
    - 15.3.2.3 Step 3 ... 234
    - 15.3.2.4 Step 4 ... 235
- 15.4 Smart Contract Design for Cloud Manufacturing ... 235
  - 15.4.1 Manufacturing Resource Verification Contract ... 235
  - 15.4.2 Manufacturing Resource Trading Contract ... 236
  - 15.4.3 The Game of Supply and Demand ... 237
  - 15.4.4 Stackelberg Balance of Resources ... 238
- 15.5 Experiment and Result Analysis ... 239
  - 15.5.1 Smart Contract Testing and Result Analysis ... 239
  - 15.5.2 Game Simulation and Result Analysis ... 240
- 15.6 Conclusion ... 242
- References ... 243

15.1 INTRODUCTION

Cloud manufacturing (Kaynak, 2020) relies on advanced technologies such as cloud computing, the Internet of Things, big data, and virtualization to provide a new production model for the manufacturing industry (Hewa et al., 2020). Based on the cloud manufacturing platform, the physical world's manufacturing resources and manufacturing capabilities are abstracted into virtual resources that can be used for consumption through digital twin technology (Leng et al., 2020).

It is still possible to break the constraints of geographical conditions. Cloud manufacturing aims to provide users with various flexible and customizable manufacturing services throughout the product life cycle (Liu et al., 2020).

As a frontier issue of common concern in academia and industry, research work on cloud manufacturing mainly focuses on reliability, resource allocation, and service quality.

Kasten et al. (2020) constructed an evaluation index system to effectively characterize cooperation trust by monitoring historical service evaluation data, quantified service satisfaction and proposed a trust evaluation model based on service satisfaction.

Bhattacharjee et al. (2020) studies the credibility of manufacturing resources in cloud manufacturing platforms and evaluates the credibility of manufacturing resources in terms of credibility and reliability. To solve the ambiguity and uncertainty of preference information in the process of matching complex product manufacturing tasks on the cloud manufacturing platform, Volpe et al. (2021) proposed a bilateral matching model based on double hesitant fuzzy preference information.

Esposito et al. (2018) offers a resource allocation method based on energy consumption perception for the energy consumption of cloud manufacturing platforms in manufacturing resource allocation. Li et al. (2019) proposes a resource bidding mechanism based on the resource scheduling problem with complex characteristics of cloud manufacturing, which ensures the fairness of the cloud manufacturing market.

Ruf et al. (2021) consider that different manufacturing cloud services have similar functions and different service qualities, a context-aware manufacturing cloud service description model is proposed to describe the dependence of a single service on other related services.

Adhikari et al. (2019) studied the impact of disturbances on service quality in the cloud manufacturing process and used discrete Markov hopping systems to achieve dynamic optimization of resources. The above-mentioned theoretical research and exploration of cloud manufacturing have made some progress in reliability, resource allocation and service quality.

However, the underlying system architecture it relies on is still a centralized framework system. Under this framework, the operator operates the cloud manufacturing platform, and the platform operator deploys both the supplier and the demander of manufacturing resources. Its most prominent feature is that the decision-making in the system depends on a small number of nodes, so it is inherently unable to avoid a single point of failure.

The single issue of failure problem is mainly solved through the redundant backup, but it requires expensive maintenance costs. It cannot fundamentally solve the single point of failure. In addition, a small number of nodes in the system have too much authority, which is easy to become the target of hacker attacks, and there is a risk of leakage of confidential data.

Blockchain technology is a new decentralized infrastructure and distributed computing paradigm (Yang et al., 2022). In the public chain, all nodes have the same status, and only when certain conditions are met can they obtain accounting rights.

Other nodes are responsible for verifying and updating the local data after the warranty. The blockchain network stimulates the enthusiasm of nodes to mine by issuing rewards. Therefore, even if there is no central node scheduling in the network, the blockchain network can still ensure the integrity and consistency of data storage.

At the same time, blockchain technology is a technology for creating trust. In a network with weak faith, nodes can be recognized by most nodes according to objective criteria, such as computing power and coinage.

It provides a secure way to exchange any goods, services or transactions. The data on the blockchain cannot be tampered with. The data stored in the block is uniquely encoded by hash operation, and the blocks are connected in a chain structure, strengthening the security of the stored data (Khang et al., AI & Blockchain, 2021).

Blockchain technology has broad application prospects. Weng Xiaoyong designed a doublechain structure to protect the shared data in the cloud platform by taking advantage of the decentralization and non-tampering properties of the blockchain (Wang et al., 2021).

Author used the data traceability of blockchain to design a food traceability system (Soni et al., 2021). To solve the expired data in the blockchain, author proposed a delectable blockchain based on an improved threshold ring signature scheme and a consensus mechanism based on proof of space (Leivadaros et al., 2021).

Xiong et al. (2020) made a systematic review of the application of blockchain in manufacturing and engineering. Wang et al. (2019) pointed out that blockchain technology can meet the needs of distributed systems with high reliability and high data security and proposed the establishment of a system platform for trusted management and control of industrial resources.

Chen et al. (2019) explores the nonzero-sum rational pricing strategy and the impact of different load levels on the benefits of all parties in the platform in a blockchain-based cloud manufacturing platform.

Shaikh et al. (2021) mainly aim at the trust problem in the cloud manufacturing platform and combine the blockchain to design a credible service transaction method. Authorities propose a distributed peer-to-peer network architecture for the centralized architecture and trust issues of third-party platforms to improve the security and scalability of the system (Shobanadevi et al., 2021).

Reference proposed a blockchain-based workflow management system to centrally share heterogeneous logistics resources of different customers. Authority offers a service composition model based on blockchain technology.

As a novel manufacturing architecture, the centralization mechanism is overcome by dividing the original service composition problem into multiple sub-problems, each containing a portion of the service/task pool.

Blockchain technology can provide an effective solution to the problems of trust and data security in cloud manufacturing systems. However, relatively few studies use the decentralization and data immutability properties of blockchain for cloud manufacturing. Therefore, this chapter proposes a blockchain-based decentralized cloud manufacturing trading platform framework.

The workload and innovation points of this chapter are as follows:

1. A blockchain-based decentralized cloud manufacturing trading platform framework is proposed. The elliptic curve digital signature algorithm in

manufacturing resources/demands is studied, and the matching method of manufacturing resources and needs.
2. The manufacturing resource verification contract and the manufacturing resource transaction contract for cloud manufacturing are designed using smart contracts, and the experimental test in the Remix platform is completed.

Explored the game problem of supply and demand balance between manufacturing resource suppliers and demanders under the decentralized architecture and conducted MATLAB simulation. The simulation results show that the game between manufacturing resource suppliers and demanders can reach Nash equilibrium and has a faster convergence rate than existing research.

15.2 BLOCKCHAIN STATEMENT

Blockchain is a decentralized distributed ledger with the advantages of common maintenance, non-tampering, openness, transparency, security and anonymity (Khang et al., IoT & Healthcare, 2021). The core components are smart contracts and consensus mechanisms, and their structure is shown in Figure 15.1.

Blockchain can be divided into public, alliance, and private chains. The public chain is a completely decentralized network. Nodes in the network have equal status and can join or leave the network at any time, represented by bitcoin and Ethereum. The consortium chain is a multi-centralized network. The consortium chain's initial members determine the number of centers.

The joining of nodes requires the approval of a specific institution, represented by Hyperledger. Finally, a private chain is a centralized network suitable for smaller groups.

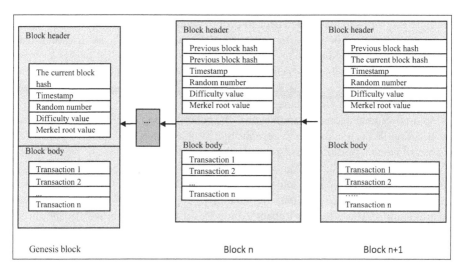

FIGURE 15.1 Blockchain structure.

A smart contract (SC) is a code embedded in hardware and can be executed automatically. Broadly speaking, a smart contract is a computerized transaction protocol that does not require an intermediary, self-verifies, and self-enforces the terms of the contract. Smart contracts endow the blockchain with greater scalability and flexibility, allowing developers to develop business logic in the blockchain network (Khang et al., AI-Centric Smart City Ecosystem, 2022).

Smart contracts use the immutability of the blockchain as the underlying support, and the entire life cycle includes contract creation, contract deployment, contract invocation, and status update. During the entire life cycle of the contract, each link is recorded in the blockchain in transactions.

Consensus algorithms are a necessary part of a blockchain system, and the consensus is the process of agreeing on data or states in a blockchain network. A blockchain system as a distributed network cannot satisfy consistency, availability and partition fault tolerance simultaneously, so a mechanism is needed to compromise between consistency and availability based on satisfying partition fault tolerance.

At present, consensus algorithms in blockchain systems can be roughly divided into proof-based and voting-based algorithms. Famous proof based consensus algorithms include workload proof, equity proof, delegated equity proof, etc. The voting-based consensus algorithm is mainly a Byzantine fault-tolerant algorithm.

15.3 BLOCKCHAIN-BASED CLOUD RESOURCE ALLOCATIONS

The traditional cloud manufacturing platform participants can be divided into manufacturing resource suppliers (manufacturing resource suppliers, MRSs), manufacturing resource demanders (manufacturing resource demanders, MRDs), cloud platform operators (cloud platform operators), platform operators, CPOs. The MRSs register the available resources with the CPOs.

The CPOs coordinate the allocation of manufacturing resources according to the needs of the MRDs, as shown in Figure 15.2, which is a centralized architecture. The blockchain-based cloud manufacturing platform proposed in this chapter can realize the allocation of manufacturing resources through smart contracts without the direct participation of a third party.

As shown in Figure 15.2, after MRSs and MRDs agree on resource price and resource supply, they sign a smart contract and save the contract in the blockchain to ensure that the contract data is not tampered with.

CPOs is responsible for overseeing the trading behavior of MRSs and MRDs and verifying the manufacturing resources used for trading by checking SCs. When the two parties have a transaction dispute, they can determine the object of the dispute through CPOs and viewing SC as Figure 15.3.

The introduction of blockchain technology can strengthen the trust between MRDs and MRSs, realize direct transactions between the two parties, and weaken the role of third parties in traditional cloud manufacturing platforms, greatly reducing the cost of credit rely

Furthermore, rely in a symmetric digital encryption and communication technology, MRSs and MRDs can grasp the usage of manufacturing resources in real-time.

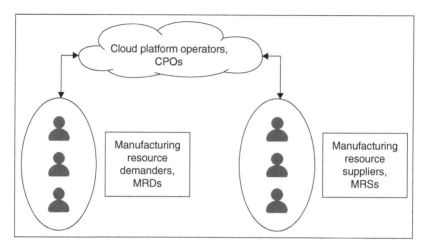

FIGURE 15.2 Traditional cloud manufacturing platform architecture.

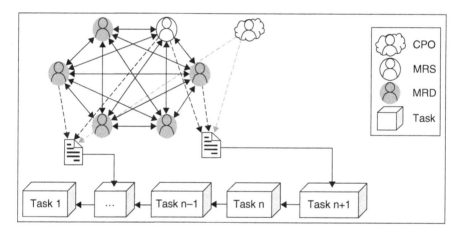

FIGURE 15.3 Blockchain-based cloud manufacturing platform architecture.

15.3.1 Manufacturing Resource/Requirement Release

To enhance the security of manufacturing resources/requirements distribution, this chapter uses an elliptic curve digital signature algorithm (ECDSA) to ensure that the data transmission between MRSs and MRDs is not tampered with. Taking the MRD sending the resource request message M to the MRS as an example, the specific steps are as follows:

MRD determines an elliptic curve E(a,b) of order p over a finite field HF(q), where a and b are curve parameters, and determines the base point H on the elliptic curve C(a,b). HF(q) = {0,1,2,..., q -1}, q is prime, and. $q \in M$. p in the following is the same as here. Without loss of generality, this chapter selects the curve $c(0,17) = y^3 - x^3 + 17 = 0$.

MRD selects a random number L_{TD} as the private key, $1 < L_{TD} < q$, and calculates the public key L_{QD}:

$$L_{QD} = L_{TD} * H \tag{15.1}$$

Select a secure hash function, perform a hash operation on the request information O, and obtain the information digest m:

$$O = SHA1(O) \tag{15.2}$$

Randomly choose an integer c, $1 < c < q$. Compute the coordinates (y_1, x_1) that map c to the elliptic curve:

$$(y_x, x_1) = c * H \tag{15.3}$$

Still get the first part s of the digital signature:

$$s = y_1 \tag{15.4}$$

If smod q =0 is the modulo operation, perform this step again. Otherwise, perform the next step.

MRD uses the private key L_{TD} to calculate another part r of the digital signature:

$$r = c^{-1}(o + s * L_{TD}) \bmod q \tag{15.5}$$

If r = 0, go back to the second step, otherwise go to the next step.

The MRD sends the resource request message O, the signature (s, r), the MSD's public key L_{QD}, the elliptic curve C(a, b) and the base point H to the MRS.

MRS verifies the received message O^* using MRD's public key L_{QD} and signature information calculate:

$$O^* = \frac{o * H + s * L_{QD}}{r} \tag{15.6}$$

If $O^* = c * H$, the verification passes, otherwise the verification fails.

15.3.2 THE MATCHING PROCESS

The matching process of manufacturing resources and manufacturing requirements, to consider the following scenario: MRD needs to produce a batch of products, and the product processing task can be decomposed into n sub-tasks Task = {$Task_1$, $Task_2$, $Task_3$,..., $Task_n$ }. Each subtask requires different processing equipment and processing time.

Therefore, MRD needs to encrypt the manufacturing demand information MMRD through the ECDSA proposed in the previous section and publish it in the blockchain network to ensure the anonymity of MRD's identity.

A hash function encrypts the address information of MRD to obtain MRD_{ad}.

$$MRD_{ad} = HASH(L_{QD}) \tag{15.7}$$

$$O_{MRD} = \{E, Q_1, U_{MRD}, MRD_{ad}, [C_{MRD}(a,b), H], L_{QD}\} \tag{15.8}$$

Among them, $E = [d_1, d_2, d_3, ..., d_n]$ represents the manufacturing resource demand of MRD's subtasks, $Q = [q_{11}, q_{12}, q_{13}, q_{1n}]$ represents the MRD a round is willing to give the purchase unit price of the manufacturing resources of each sub-task, $U_{MRD} = \{U_{MRD}, U_{MRD1}, U_{MRD2}, U_{MRD3}, U_{MRDn}\}$ represents the increase of each sub-task

Working time period, MRD_{ad} represents the address information of $[C_{MRD}(a,b), H]$ represents the information data used by MRD for signature, $C_{MRD}(a,b)$ represents elliptic curve, H is the base point; L_{QD}

15.3.2.1 Step 1
MRD uses ECDSA to digitally sign the manufacturing requirement information O and broadcast the signature and the required information to the network.

15.3.2.2 Step 2
After receiving the manufacturing resource request message M'_{MRD} from MRD, MRS first verifies the message's validity and verifies whether the message is a message sent by MRD through the public key L_{QD} provided by MRD. If the message is invalid, it will not respond; if the message is valid, the MRS will check the request information M'_{MRD}, and reply to the manufacturing resource information O_{MRS} that can be provided to the MRD to the MRD.

$$O_{MRS} = \{R, P_1, U_{MRS}, MRS_{ad}, [C_{MRD}(a,b), H], L_{QD}\} \tag{15.9}$$

Where $R = \{R_1, R_2, R_3, ..., R_n\}$ represents the manufacturing resources that MRS can provide, $P = \{p_{11}, p_{12}, p_{13}, ..., p_{1n}\}$ represents the number of resources that MRS can provide for each requested resource. The unit price of one round of selling, $U_{MRD} = \{U_{MRD}, U_{MRD1}, U_{MRD2}, U_{MRD3}, U_{MRDn}\}$ represents the available time period of the resource.

15.3.2.3 Step 3
After receiving the manufacturing resource supply message "MMRS" from MRS, MRD first verifies the message's validity and verifies whether the message is a message sent by MRS through the public key KPS provided by MRS. If the message is invalid, it will not respond; if the message is valid, MRD will check the reply message 'MMRS," and negotiate the disputed place, such as price, to send the message to MRS.

$$O_{MRS} = \{Q_2, MRS_{ad}, [C_E(a,b), H], L_{QD}\} \tag{15.10}$$

$Q_2 = \{q_{21}, q_{22}, q_{23}, ..., q_{2n}\}$ represents the second round of offers for MRD. The price negotiation may last for several rounds, and the game between MRS and MRD will eventually reach an equilibrium point. If both parties can accept the price at the equilibrium moment, the transaction reaches a consensus and proceeds to the next step; if at least one party cannot accept the price at the equilibrium point, the transaction fails, and MRD re-publishes the resource request message to the network.

15.3.2.4 Step 4

Both MRS and MRD sign the smart contract for the transaction. In the process of matching manufacturing resources and demand, according to the price, both parties to the transaction independently complete the matching and transaction. After the contract is signed, it is stored in the blockchain to ensure that the transaction record is not tampered.

15.4 SMART CONTRACT DESIGN FOR CLOUD MANUFACTURING

Smart contracts are an important part of the blockchain. They are stored in a specific location on the blockchain and can be called and automatically executed by other nodes, giving the blockchain the characteristics of intelligence. Due to the openness and transparency of the blockchain, all nodes can judge the output of the contract according to the input before running the contract formally, so there is no fraud through smart contracts.

Despite the advantages of smart contracts, few studies have applied this advantage to actual production. Literature combined smart contracts with data in the Industrial Internet of Things, studied data package contracts and data analysis service contracts, and realized the transaction of data commodities. Inspired by this, this chapter proposes a blockchain-based cloud manufacturing framework.

In the process of resource allocation, the security and credibility of transactions are guaranteed by signing manufacturing resource verification contracts and manufacturing resource transaction contracts.

15.4.1 MANUFACTURING RESOURCE VERIFICATION CONTRACT

The manufacturing resources stored in the cloud manufacturing resource pool are the digital version of the manufacturing resources in the physical world. In the process of virtualization, to obtain greater benefits, MRSs may falsify virtual resources by some means, such as deliberately exaggerating the number of resources.

Although the blockchain can guarantee the immutability of the data on the chain, it cannot guarantee the authenticity of the original data in uploading the data to the chain. Therefore, the introduction of a regulatory mechanism is critical. Therefore, when introducing blockchain technology, this chapter does not completely abandon the supervision function of CPOs due to its decentralization characteristics. The supervision function of CPOs only exists in the first uploading of manufacturing resources.

This chapter mainly relies on CPOs and SC to complete the verification of data uploading. If the manufacturing resources reported by MRSs exceed the range of manufacturing resources counted by CPOs, the reporting process will be intercepted to improve the data reliability of the system.

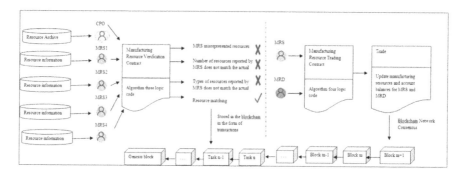

FIGURE 15.4 Smart contract design process.

Once the data of the manufacturing resource is uploaded to the blockchain network, the system will judge the legality of the manufacturing resource based on the historical records. Therefore, retaining the supervision function of CPOs will increase the reliability of the data, although it will add a verification link and reduce the efficiency of data uploading as Figure 15.4.

The manufacturing resources stored in the cloud manufacturing resource pool are the digital version of the manufacturing resources in the physical world. In the process of virtualization, to obtain greater benefits, MRSs may falsify virtual resources by some means, such as deliberately exaggerating the number of resources.

Although the blockchain can guarantee the immutability of the data on the chain, it cannot guarantee the authenticity of the original data in uploading the data to the chain.

Therefore, the introduction of a regulatory mechanism is critical. Therefore, when introducing blockchain technology, this chapter does not completely abandon the supervision function of CPOs due to its decentralization characteristics. However, the supervision function of CPOs only exists in the first uploading of manufacturing resources. Therefore, this chapter mainly relies on CPOs and SC to complete the verification of data uploading. If the manufacturing resources reported by MRSs exceed the range of manufacturing resources counted by CPOs, the reporting process will be intercepted to improve the data reliability of the system.

Once the data of the manufacturing resource is uploaded to the blockchain network, the system will judge the legality of the manufacturing resource based on the historical records. Therefore, retaining the supervision function of CPOs will increase the reliability of the data, although it will add a verification link and reduce the efficiency of data uploading.

15.4.2 Manufacturing Resource Trading Contract

After MRS and MRD reach a consensus on the resource price, type and quantity through the blockchain communication network, the authenticity and validity of the transaction are guaranteed by signing a smart contract. The price Q in the contract is the price agreed upon by both parties. The MRS displays the type r_i.name and

$r_i \cdot num, r_i \in R$ of all manufacturing resources R in the contract. MRD selects the relevant manufacturing resource type d j .name and quantity $e_j \cdot num, e_j \in E$ according to the purchase demand E. During the transaction process, the manufacturing resource smart contract calculates the payment amount according to the MRD demand; MRS supply and resource price, and sends the amount to the MRS account balance [MRS].

At the same time, adjust the number of resources corresponding to MRS and the account balance [MRD], as shown in Figure 15.4. Once the smart contract runs on the blockchain, all transaction records will be permanently stored and cannot be tampered with, thus ensuring the authenticity and reliability of the transaction.

15.4.3 THE GAME OF SUPPLY AND DEMAND

The game of supply and demand of manufacturing resources are the blockchain-based cloud manufacturing resource allocation framework enables MRSs and MRDs to communicate in real-time and bi-directionally through the blockchain network and enables both parties to keep abreast of each other's manufacturing resource demand/supply situation. Therefore, MRDs can determine the purchase amount of resources according to the unit resource price and supply provided by MRSs and their own demand. MRSs can also adjust the supply of manufacturing resources and resource price according to the resource demand of MRDs.

There are m MRDs and 1 MRS in the cloud manufacturing platform for a certain manufacturing resource S. Each MRD publishes the demand du of resource S through the blockchain network. MRS determines the supply W_s of the resource after summarizing all demand information and can provide a sufficient amount of resource S to obtain the optimal income.

Therefore, the two parties have a time sequence when making decisions, which is a dynamic process with complete information. Therefore, this chapter adopts the related theory of the Stackelberg game to solve the problem of resource income between MRDs and MRS, in which MRDs is the leader, MRS is the follower, and MRS determines the resource supply according to the resource demand of MRDs, forming a multi-leader-follower issuer.

First, MRDs collect and publish the manufacturing resource demand V_d, $V_d = \{v_{d1}, v_{d2}, v_{d3}, ..., v_{dm}\}$ through the blockchain network. v_{dj} is the resource requirement of each MRD, $j \in \{1,2,...,m\}$. After MRDs obtain manufacturing resources, they can benefit by processing and producing products.

The benefit of converting manufacturing resources into product benefits is Y, $Y = \{y_1, y_2, y_3, .., y_m\}$ denotes the benefit of each MRD, $j \in \{1,2,...,m\}$. The price paid per unit of manufacturing resource S is $Q_s = [q_s]$, and the production cost D_d, $D_d = [c_{d1}, c_{d2}, c_{d3}, ..., c_{dm}]$ represents the production cost of each MRD, $j \in \{1,2,...,m\}$. Therefore, the profit of MRDs is:

$$I = V_d \{Y - Q_s - D_d\} \tag{15.11}$$

After MRD understands the manufacturing demand, it determines the supply of manufacturing resources $W_s = [w_s]$. The price of manufacturing resources is affected by supply and demand, which is positively correlated with demand and negatively correlated with supply. Therefore, the resource selling price is defined as:

$$Q_s = b\sum_{j=1}^{m} v_{dj} - cv_s \qquad (15.12)$$

b>0, c>0 is the influence coefficient of demand and supply on price. The MRS is responsible for the routine maintenance of the manufacturing resource S, and the maintenance cost per unit resource is recorded as $D_s = [d_s]$ So the profit for MRS is:

$$G = W_s(Q_s - D_s) \qquad (15.13)$$

The purpose of the game is to maximize the profit for the participants. Therefore, the objective function is:

$$\begin{cases} \max_{v_{dj}} I \\ \max_{w_s} G \\ s.t. v_{dj} \geq 0, w_s > 0 \end{cases} \qquad (15.14)$$

15.4.4 Stackelberg Balance of Resources

The Stackelberg game is the dominant game. In this chapter, MRDs are the dominant players, and the resource demand is determined first, and then the MRS determines the supply according to the demand.

The ultimate goal of both sides of the game is to gradually adjust their own strategies under the constraints of the other party's strategies to maximize their own interests. When the interests are maximized, the strategy sets of both parties will reach a relatively stable state; that is, the Nash equilibrium will be reached.

Lemma 1 For the resource price $\alpha=[\alpha_1,\alpha_2,\alpha_3,\ldots,\alpha_k,\ldots,\alpha_m]$, m is the number of players in the game, and the difference between it and the profit $\beta(\alpha)$ Stackelberg game, if conditions I and II are satisfied, there is a Nash equilibrium.

The condition I: α is a nonempty bounded closed convex subset on Euclidean space.

Condition II: $\forall \beta_j \in \beta, \beta_j$ is continuous and concave concerning α_j.

Theorem 1 For the Stackelberg game of supply and demand of manufacturing resources described by equations (15.11)-(15.14), there is the Nash equilibrium.

Proof because for $\forall j \in \{1,2,..m\}, v_{dj} > 0 w_s > 0$. So the policy set is a nonempty bounded closed convex subset on Euclidean space. So condition (I) is established also because:

$$i_j = y_j v_{dj} - \left(b\sum_{j=1}^{m} v_{dj} - cw_s + d_{dj}\right) \qquad (15.15)$$

$$g = w_s\left(b\sum_{j=1}^{m} v_{dj} - d_s\right) - cw_s^2 \qquad (15.16)$$

It is known from equations (15) and (16) that i_j is continuous concerning v_{dj}, and g is continuous concerning $v_{dj} W_s$.

$$\frac{e^2 i_j}{e v_{dj}^2} = -2a < 0, \forall j \in \{1,2,..m\} \quad (15.17)$$

$$\frac{e^2 g}{e w_s^2} = -2b < 0 \quad (15.18)$$

It is known from equations (15.17) and (15.18) that i_j is concave concerning v_{dj}, and g is concave concerning W_s. So condition (II) holds.

Using the reverse induction method, assuming that MRS has reached the equilibrium point, resource supply $W_s = W_s^* = [W_s^*]$ substitute into equation (15.15), take the derivative of equation (15.15) concerning, v_{dj} and set the derivative function equal to zero to get:

$$v_{dj}^* = \frac{y_i - D_{dj} + CW_s^* - b\sum_{i=1, i \neq j}^{m} v_{dj}}{2b} \quad (15.19)$$

Substitute Equation (15.19) into Equation (15.16), and derive Equation (15.16) concerning w_s, and set the derivative function equal to zero to obtain.

15.5 EXPERIMENT AND RESULT ANALYSIS

15.5.1 SMART CONTRACT TESTING AND RESULT ANALYSIS

The test of the smart contract uses Remix as the test environment. A remix is a browser-based compiler and IDE that allows users to build Ethereum contracts and debug transactions using the solidity language.

In the manufacturing resource verification contract, MRS and CPO upload manufacturing resource information in the contract, respectively. Then, deploy the contract into the Ethereum blockchain network.

To verify the contract's validity, four tests are performed on the manufacturing resource on-chain verification contract, and the test input data is shown in Table 15.1.

In Test 1, the types of resources reported by the MRS were consistent with the types assessed by the CPO, and the number of resources was within the scope of the assessment. Therefore, the resource information is successfully uploaded to the chain, a transaction is created in the blockchain network, and the hash value of the transaction is generated.

In Test 2, MRS maliciously reported the type and quantity of resources, in Test 3, MRS exaggerated the number of resources and in Test 4, MRS falsely reported the type of resources, which were intercepted by the blockchain network, interrupted the execution of the transaction, and rolled back to the chain.

To maintain the authenticity and credibility of the data in the cloud manufacturing platform, the original data that has been tampered with is kept on the chain. Therefore,

TABLE 15.1
Input Data of Contract Test

	MRS resource name	MRS resource quantity	CPO funding source name	CPO funding number of sources
Test 1	m1,m2	10,20	m1,m2	10,30
Test 2	m1,m2,m3	10,20,30	m1,m2	10,30
Test 3	m1,m2	20,30	m1,m2	10,30
Test 3	m1,m3	10,20	m1,m2	10,30

TABLE 15.2
Comparison over Cloud Parameter

Evaluation Parameter	Speed of Game			
	MRD1	MRD2	MRD3	MRS
Converge speed	52	54	50	45
Load overhead	48	45	55	59
Converge cost	58	60	65	70
Supply demand balance	78	82	71	65

the blockchain-based cloud platform architecture can effectively supervise the operation records in the platform, facilitate tracing the origin of disputes, and maintain the openness and transparency of the platform.

15.5.2 GAME SIMULATION AND RESULT ANALYSIS

Under the cloud manufacturing resource allocation framework based on blockchain, resource prices are affected by supply and demand, MRDs lease relevant manufacturing resources according to production needs, and MRS formulates relevant supply strategies according to resource demand maximum profit under the constraints of the strategy (Pankaj Bhambri et al., Cloud & IoT, 2022).

The experiment uses MATLAB2019b to carry out the real supply and demand game between 3 MRDs and 1 MRS. The coefficients of the price function are a = 0.8, b = 0.2.

This chapter assumes that the impact of price demand is greater than that of supply as Table 15.2.

Y =[100,200,200], D_d =[45,45,15], D_s =[10]. The initial demand V_d = [100,100,100] and the supply W_s=[200]. The manufacturing resource game model proposed in this chapter is compared with the power grid supply and demand game model proposed in.

Although the research objects are different, both study the problem of Stackelberg equilibrium convergence, so there is certain comparative research significance.

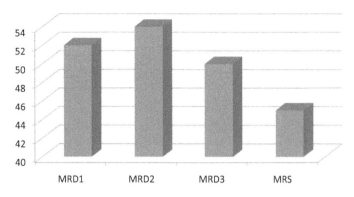

FIGURE 15.5 Convergence speed.

Source: Chart of Convergence Speed is designed by Akhilesh Kumar, Nihar Ranjan Nayak, Samrat Ray, and Ashish Kumar Tamrakar.

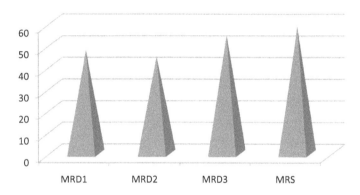

FIGURE 15.6 Load overhead.

According to the parameter settings in the literature, this chapter makes a real comparison in MATLAB 2019b as Figure 15.6.

It can be seen from Figure 15.6 that this chapter is close to convergence after about 15 iterations. In comparison, the literature is close to convergence after 35 iterations, and the convergence speed is increased by 57% as Figure 15.7.

About Figures 15.5–15.8 the impact of the increase in MRD3 demand on the balance of supply and demand after the balance of supply and demand, MRD3 privately increased the demand to obtain higher returns, breaking the balance of supply and demand.

The increase in demand led to increased resource prices, which brought a greater impact to MRS. Profit. It also increases the purchase cost of MRD3 and MRD2. Therefore, MRS gains more profit by hoping to increase the supply, while MRD2 and MRD3 are forced to reduce resource demand. Still, the price is balanced.

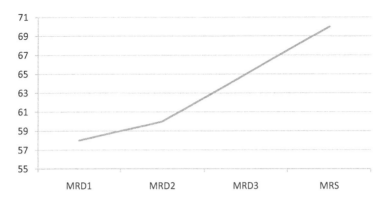

FIGURE 15.7 Convergence cost.

FIGURE 15.8 Supply demand balance.

Source: Chart of Supply Demand Balance is designed by Akhilesh Kumar, Nihar Ranjan Nayak, Samrat Ray, Ashish Kumar Tamrakar.

Ultimately 3 1 MRDs and 1 MRS return to the original supply and demand balance, as shown in Figures 15.5–15.8.

15.6 CONCLUSION

This chapter mainly studies the application of the integration of blockchain technology and cloud manufacturing platforms and proposes a blockchain-based cloud manufacturing resource allocation framework. At the same time, the Stackelberg supply and demand game problem between the resource demander and the supply side under the decentralized framework is studied.

Under this framework, the participants of the cloud platform use the elliptic curve digital signature algorithm to complete the distribution of resources through the blockchain network and complete the matching and transaction of resources through smart contracts.

Through the test of intelligent contracts through Remix, the results show that the transaction data of the blockchain-based cloud platform can be safely stored in the blockchain, and the immutability of the blockchain can enhance the credibility of cloud manufacturing data.

In the multi-leader-follower game model, the manufacturing resource supplier and the resource demander can obtain the Nash equilibrium. The results show that this model's Nash equilibrium convergence speed is greatly improved compared with the existing research, and it has certain robustness.

In the future, we will study the optimization of the consensus algorithm of blockchain, reduce its consensus loss, and explore how to improve the throughput of its application in the cloud manufacturing platform.

REFERENCES

Adhikari, A., & M. Winslett, A hybrid architecture for secure management of manufacturing data in Industry 4.0, *2019 IEEE International Conference on Pervasive Computing and Communications Workshops (PerCom Workshops)*, 2019, pp. 973–978, doi: 10.1109/PERCOMW.2019.8730717.

Bhambri, Khang, A., Sita Rani, & Gaurav Gupta, *Cloud and Fog Computing Platforms for Internet of Things*. Chapman & Hall. ISBN: 978-1-032-101507, 2022, https://doi.org/10.1201/9781032101507.

Bhattacharjee, A., S. Badsha, & S. Sengupta, Blockchain-based secure and reliable manufacturing system, *2020 International Conferences on Internet of Things (iThings) and IEEE Green Computing and Communications (GreenCom) and IEEE Cyber, Physical and Social Computing (CPSCom) and IEEE Smart Data (SmartData) and IEEE Congress on Cybermatics (Cybermatics)*, 2020, pp. 228–33, doi: 10.1109/iThingsGreenCom-CPSCom-SmartData-Cybermatics50389.2020.00052.

Chen, Q.. Q. Xu, & C. Wu, Optimal sharing strategies of idle manufacturing resource considering the effect of supply-demand matching, 2019 *International Conference on Industrial Engineering and Systems Management (IESM)*, 2019, pp. 1–6, doi: 10.1109/IESM45758.2019.8948199

Esposito, C., F. Palmieri, & K.-K.R. Choo, Cloud message queueing and notification: challenges and opportunities. *IEEE Cloud Computing* 5(2), 11–16, Mar./Apr. 2018, doi: 10.1109/MCC.2018.022171662

Hewa, T. M., A. Braeken, M. Liyanage, & M. Ylianttila, Fog computing and blockchain based security service architecture for 5G industrial IoT enabled cloud manufacturing, *IEEE Transactions on Industrial Informatics*, doi: 10.1109/TII.2022.3140792

Kasten, J. E. Engineering and manufacturing on the blockchain: A systematic review, *IEEE Engineering Management Review* 48(1), 31–47, March 2020, doi: 10.1109/EMR.2020.2964224

Kaynak, B., S. Kaynak, & Ö. Uygun, Cloud manufacturing architecture based on public blockchain technology. *IEEE Access*. 8, 2163–77, 2020, doi: 10.1109/ACCESS.2019.2962232

Khang, A., Rani, S., & Sivaraman, A.K. (Eds.). *AI-Centric Smart City Ecosystems: Technologies, Design and Implementation* (1st ed.). CRC Press, 2022. https://doi.org/10.1201/9781003252542

Leivadaros, S., G. Kornaros, & M. Coppola, Secure asset tracking in manufacturing through employing IOTA distributed ledger technology, *2021 IEEE/ACM 21st International*

Symposium on Cluster, Cloud and Internet Computing (CCGrid), 2021, pp. 754–61, doi: 10.1109/CCGrid51090.2021.00091

Leng, J. et al., ManuChain: Combining permissioned blockchain with a holistic optimization model as bi-level intelligence for smart manufacturing, *IEEE Transactions on Systems, Man, and Cybernetics: Systems* 50(1), 182–92, Jan. 2020, doi: 10.1109/TSMC.2019.2930418

Li, R., T. Chen, P. Lou, J. Yan, & J.Hu, Trust mechanism of cloud manufacturing service platform based on blockchain, *2019 11th International Conference on Intelligent Human-Machine Systems and Cybernetics (IHMSC)*, 2019, 15–19, doi: 10.1109/IHMSC.2019.10099

Liu, Y., J. Zhang, L. Zhang, & H. Liang, IoT – and blockchain-enabled credible scheduling in cloud manufacturing: a systemic framework, *2020 IEEE 18th International Conference on Industrial Informatics (INDIN)*, 2020, 488–93, doi: 10.1109/INDIN45582.2020.9442088

Rana, Geeta, Alex Khang, Ravindra Sharma, Alok Kumar Goel, & Ashok Kumar Dubey, The role of artificial intelligence in blockchain applications, *Reinventing Manufacturing and Business Processes through Artificial Intelligence*, 2021, doi: 10.1201/9781003145011

Rani, Sita, Khang, A., Meetali Chauhan, & Aman Kataria, IoT equipped intelligent distributed framework for smart healthcare systems, *Networking and Internet Architecture*, 2021, https://arxiv.org/abs/2110.04997v2, doi: 10.48550/arXiv.2110.04997

Ruf, P., J. Stodt, & C. Reich, Security threats of a blockchain-based platform for industry ecosystems in the cloud, *2021 Fifth World Conference on Smart Trends in Systems Security and Sustainability (WorldS4)*, 2021, 192–9, doi: 10.1109/WorldS451998.2021.9514058

Shaikh, E., A. Bashar, & N. Mohammad, Recent applications of computing and mobility technologies to modern manufacturing, *2020 International Conference on Communications, Signal Processing, and their Applications (ICCSPA)*, 2021, 1–6, doi: 10.1109/ICCSPA49915.2021.9385756

Shobanadevi, A., Tharewal, S., Soni, M. et al. Novel identity management system using smart blockchain technology. *Int J Syst Assur Eng Manag*, 2021, https://doi.org/10.1007/s13198-021-01494-0

Soni, M. & D. K.Singh, Blockchain implementation for privacy preserving and securing the healthcare data, *2021 10th IEEE International Conference on Communication Systems and Network Technologies (CSNT)*, 2021, 729–34, doi: 10.1109/CSNT51715.2021.9509722

Volpe, G., A. M. Mangini, & M. P. Fanti, An architecture for digital processes in manufacturing with blockchain, docker and cloud storage, *2021 IEEE 17th International Conference on Automation Science and Engineering (CASE)*, 2021, pp. 3944, doi: 10.1109/CASE49439.2021.9551633

Wang, M., C. Xu, X. Chen, L. Zhong, Z. Wu, & D.O. Wu, BC-mobile device cloud: A blockchain-based decentralized truthful framework for mobile device cloud, *IEEE Transactions on Industrial Informatics*, 17(2), 1208–19, Feb. 2021, doi: 10.1109/TII.2020.2983209

Wang, Y., X. Sun, F. Zhu, F. Zhang, M. Zhang, & H. Cao, Chain FileSynch: An innovate file synchronization for cloud storage with blockchain, *2019 International Conference on Artificial Intelligence and Advanced Manufacturing (AIAM)*, 2019, 552–6, doi: 10.1109/AIAM48774.2019.00115

Xiong, Z., J. Kang, D. Niyato, P. Wang, & H.V. Poor, cloud/edge computing service management in blockchain networks: Multi-leader multi-follower game-based ADMM for pricing, *IEEE Transactions on Services Computing* 13(2), 356–67, March 1–April 2020, doi: 10.1109/TSC.2019.2947914

Yang, C., S. Lan, Z. Zhao, M. Zhang, W. Wu, & G.Q. Huang, edge-cloud blockchain and IoE enabled quality management platform for perishable supply chain logistics, *IEEE Internet of Things Journal*, doi: 10.1109/JIOT.2022.3142095

16 Blockchain-based Privacy Protection Credential Model for Zero-Knowledge Proof over Distributed Systems

*Sarfraz Fayaz Khan, Sumit Kumar,
Ramya Govindaraj, and Sagar Dhanraj Pande*

CONTENTS

16.1	Introduction	248
16.2	Related Work	250
16.3	System Model	251
	16.3.1 System Goals	251
	16.3.1.1 Autonomous Control	251
	16.3.1.2 Distributed Authentication	251
	16.3.1.3 Build Trust	251
	16.3.1.4 Privacy Protection	251
	16.3.1.5 Revocability	252
	16.3.1.6 Portability	252
	16.3.2 System Architecture	252
16.4	Blockchain-Based Distributed Identity Authentication System	253
	16.4.1 Symbol Definition	253
	16.4.1.1 DID	253
	16.4.1.2 Subject	253
	16.4.1.3 Claim	254
	16.4.1.4 Verifiable Credential	254
	16.4.1.5 Presentation	254
	16.4.2 Algorithm Description	254
	16.4.2.1 Digital Identity Management System	254
	16.4.2.2 Credential Management System	255
	16.4.3 System Application	259
16.5	Experimental Evaluation	259
16.6	Smart Contract Design	259

16.7 System Performance Test 260
16.8 Conclusion 264
References 264

16.1 INTRODUCTION

With the continuous change of the microgrid industry and the distribution of new energy, the bureau occupies an important strategic position in the China (Tu et al., 2019). On April 2020, "On Doing a Good Job in Renewable Energy" issued by the General Department of the National Energy Administration Communication on matters related to the preparation of the "14th Five-Year Plan" of Yuanyuan Development Knowing, the notice stated: Many governments have been prioritizing the development of local decentralized and distributed renewable energy resources, vigorously promote distributed renewable energy Power, heat, gas, etc. are used directly and nearby on the user side.

Integrate new technologies such as energy storage and hydrogen energy to increase the use of renewable energy in the region share of the region's energy supply. "Distributed renewable energy" is the future development direction, while developing in such a distributed system Charging stations and even charging and discharging stations are undoubtedly the general trend, and there are even May become standard (Shobana et al., 2019).

With the wide application of new energy such as wind power, solar the proportion can be greatly increased. For example, vigorously developing optical storage and charging integrated microgrid technology to supply clean energy through photovoltaics. After electricity is generated, electricity is stored, and photovoltaic energy storage and charging facilities are formed developing a microgrid that interacts intelligently with the public grid based on demand. It can realize two different operation modes of grid-connected and off-grid (Zou et al., 2021).

To alleviate the impact on the regional power grid when the charging pile is charged with high current the transmission and distribution process of the energy storage system is usually accompanied by distributed generations such as electricity, distributed electricity sales, smart meters, terminal smart charging piles, and other service needs. The rapid expansion of micro-grids has also brought more and more stringent serious security challenges (Soni et al., 2021).

New power supply based on wind power and photovoltaic is the system presents a distributed architecture, and the centralized user management system is unable to meet the point-to-point trusted communication between distributed subjects as mutual needs in article (Z. Wu et al, 2020). In addition, the terminal nodes of the energy Internet of Things are widely distributed, wide range, large number, complex environment and limited computing resources, extremely It is vulnerable to attacks such as fraudulent use and tampering (S. Khan et al, 2020). In addition, the terminal nodes of the energy Internet of Things are widely distributed, Wide range, large number, complex environment and limited computing resources, extremely It is vulnerable to attacks such as fraudulent use and tampering (Khan et al., 2020).

In order to meet the new power development requirements, in 2019, the State Grid The State Grid company proposed to "carry out research on the application of new technologies such as blockchain, effectively supporting and promoting the integrated development of 'two networks', promoting blockchain technology and furthering research and deepening application of technology in the power industry (Yan et al., 2021).

The technical form of chain opening, sharing and collaboration and the construction of State Grid Corporation Internationally leading microgrids with Chinese characteristics are highly compatible with the strategic goals and can In order to effectively solve the data fusion in the process of ubiquitous microgrid construction communication, device security, personal privacy, architectural rigidity, and multi-agent collaboration The same problems are seen in the construction of ubiquitous microgrids as description in the role of generation (Shobanadevi et al., 2021). Blockchain technology is characterized by its decentralization, security and transparency, and it is not easy to Characteristics such as tampering and the demand for distributed energy transactions are extremely high (Boneh et al., 2001).

In order to solve the identity management of terminal nodes in the micro-grid management, and trusted access as well as business access control, this section is based on FISCO BCOS consortium blockchain technology, designed a multi-center A smart distributed identity authentication system that supports fine-grained statement descriptions (Khang et al., AI & Blockchain, 2021).

Describes and privacy credentials based on zero-knowledge proofs, realizing entity identity self-control of shares, fine-grained access control at attribute level, user identity privacy protection and trusted data exchange. Specifically, the contributions are as follows:

a. Autonomous control of user identity: in contrast to traditional identity management identity no longer has to be in the hands of the identity provider, since users fully own, control, and manage their own identity.
b. Distributed trusted access: we propose a consortium-based blockchain distributed identity management technology that supports flattened distributed identity management certificate, and does not depend on the certificate of a single enterprise, on the certificate of a single enterprise through the distributed application (Khang et al., IoT & Healthcare, 2021).
c. In the social relations of households have obtained comprehensive identity authentication in the era of digital transformation, which is truly real in the most developed countries. Now the user has identity autonomy, through KYC (know your customer). The technology realizes identity supervision and solves the problem of micro-grid terminal nodes.
d. Verifiable credentials for microgrids are based on fine-grained attributes of users and based on zero-knowledge proof technology, the system will supports the generation of anonymous credentials. For sensitive information of users, it is possible to disclose confidential information without disclosing in the case of complete verification, the user's identity is effectively hidden.

Based on blockchain technology, it supports multiple certificate authorities and corresponds to different trust levels. It is a distributed multi-center identity management platform that supports multi-level migration of identities and credentials between platforms, chains, and applications.

16.2 RELATED WORK

With the popularization and development of the Internet, network digital identity came into being, and it used to fill in the missing identity layer of the Internet and solve the problem of trusted interactions.

The first generation of network identities is the application accounts that assigned a local account and simply withdraws from a centralized identity provider by using username and account password. Because local accounts between different applications cannot communicate, people need to maintain a large number of application accounts, based on the alliance account identity produced. SSO, single sign-on system (Yong et al., 2016), is a popular alliance body.

A real-time data synchronization solution that can address access across multiple application systems, alliance all application systems and trust issues across systems. Traditionally, the authentication system of the microgrid is generally also through the SSO Chen et al. (2011) established a coalition in a heterogeneous environment. Their architecture proposes a model based on IdP/SP and SAML technology authentication solutions (Hong, Z., 2014).

The State Grid Management has built an identity management and authentication system with a unified index of the whole network system, providing identity authentication, accounts for the major business systems, SSO and other infrastructure services, saving storage overhead and reducing individual data retention between sub-application systems.

The security risk is avoided, the dilemma of data silos is avoided, and the basic solution is the interconnection of the whole network. However, in a centralized system the user's identity information is completely controlled by the service provider and may exist activities of "Information Excessive Collection", "Malicious Collection Analysis", "Data Breach and Data Uncontrollable Behavior" such as buying and selling.

In order to achieve user identity autonomy, researchers began to explore blockchain-based digital identity management schemes (Niu & Li, 2014). A public key base (PKI) (Wang et al., 2016) is a commonly used identity and certificate management technology.

Lu et al. (2018) designed and developed a blockchain-based PKI management system for issuing, validating, and revoking X.509 certificates to and addressing single points of failure and enabling rapid response to CA-centric defects.

uPort (Zhou et al., 2020) is a development based on Ethereum and distributed network design. It is an open identity system, based on smart contracts to support multi-center the autonomy of identity data and supports multi-centralized applications.

Zhao et al. (2019 presents an identity management authority to individuals and implements it in the blockchain system shared, enabling interaction between multiple systems and applications and services.

Blockchain-based identity authentication systems are mostly based on the public blockchain platform; there is no alliance scenario for microgrid effective transformation and integration, and it is difficult to meet the requirements of trusted access, fine-grained access control, and other requirements, so it is not suitable for grid scenarios (Rani et al., 2021).

Although there are many applications of the blockchain in the power grid, most of them are centralized due to the application of the blockchain platform in the grid transaction scenario, and the lack of solutions to distributed identity trustworthiness from the perspective of identity management. Therefore, this chapter is based on the FISCO BCOS consortium blockchain technology, and proposes a distributed identity that supports multi-center authentication to realize autonomous control and trusted connection of user identity input, and achieve attribute-level fine-grained through attribute declaration technology (Kataria et al., 2022).

16.3 SYSTEM MODEL

16.3.1 System Goals

Trusted identity authentication as the underlying infrastructure of identity governance implementation to solve cross-department, cross-organization, and cross-environment. Trusted identity authentication problem to ensure that the user's identity information is in their own hands control, protect privacy and personal information, while retaining inspections for regulatory purposes verification channel (such as KYC verification or connection with eID). Its goals are as follows.

16.3.1.1 Autonomous Control
DID can be established by users themselves, and users fully own, control, and manage their own identities.

16.3.1.2 Distributed Authentication
It does not depend on the authenticatioIt does not depend on the It does not depend on the authentication of a single enterprise, but through distributed application. In this way, the social relations of households obtain comprehensive identity authentication.

16.3.1.3 Build Trust
This chapter first selects the committee node as the issuer and accounting section point, and blockchain technology is designed to ensure that every accounting node in the application network and all connected points are involved in data interaction and record management, etc., and then according to KYC's real user information records and audit rules. The information provided is reviewed and the committee node is used as a credential. The issuer endorses the user credentials, which can realize a multi-center machine. The establishment of trust bridges in the structure is included, thereby realizing the characteristics of supporting multi-center mutual trust between peers and nodes.

16.3.1.4 Privacy Protection
The user's digital identity identifier cannot infer real user identity information, and the trusted authentication identifier and verifiable digital certificate of a user are stored off-chain and stored on-chain. Moreover, user attribute table shown as a discrete

logarithmic model, converted to password commitment hidden user hidden (Zhao et al., 2020) privacy, to achieve a higher level of privacy requirements.

16.3.1.5 Revocability

Use the revocation strategy supported by the auxiliary chain to realize the password credential revocation inquiry, providing proof of non-revocation.

16.3.1.6 Portability

Support multi-platform, cross-chain, cross-application identification and certified credentials transplant.

16.3.2 System Architecture

Similar to the traditional PKI system, in the identity authentication ecosystem, the system has four roles, namely: Issuer, Holder, Verifier (Verifier), and Identifier Registry, where the system can have multiple certification centers, as shown in Figure 16.1.

In addition, institutional committee members and system administrators are also alliance chain roles, with privileges and system administration privileges for the access license, system role tool. The diagram is described below. Issuer: Possesses user data and can issue verifiable claims. Holder can request, stores or sends claims to verifies. Verifier is credentials entities such as governments, banks, universities, etc. in the blockchain. The committee acts as the issuer. User-specific identity social attributes (Khang et al., AI-Centric Smart City Ecosystem, 2022).

The contradiction between diversity and unified identity management center, this section adopts to replace the traditional centralized certificate center with multi-authority issuance. Assumption, there is already a licensing model for selecting committee nodes, the system select the committee (Identifier Registry) as the issuer of the certificate as Figure 16.1.

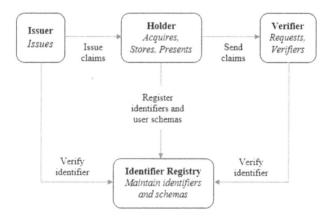

FIGURE 16.1 Architecture of identification system.

Let C_{mt} denote the commission members,

Committee nodes Self-stores a private key sk for issuing credentials. The corresponding public keypk is used to verify credentials. Either party (distributed application or committee nodes) can act as credential issuers, the issuer's. The main functions include issuer registration/deregistration, issuing certificate templates, issuing user credentials, etc.

Assisted by KYC rules, the information provided is reviewed and the committee node is used as a credential, and the issuer endorses the user credentials to establish a multi-center organization.

A bridge of trust is authenticator, it also known as the application side, accepts verifiable claims credentials and validate so that it can be presented to the person presenting the verifiable claim credential some type of service. The main functions of a validator include publishing validation rules, verifying user DID through on-chain information, and verifying user credentials.

Validators include employers, security personnel, and websites, among others. Holder: Requesting Verifiable Claims Credentials from Issuer entity. Present the verifiable claim credential to the verifier. The main functions of the holder include DID registration/cancellation/renewal, applying for a certificate to the issuer, validators present credentials, etc. Holders include students, staff and customers, etc.

Identifier Registry: Responsible for maintaining the database of DIDs, such as a blockchain, distributed ledger (e.g., in DID field), which is convenient for the verifier to verify the claim credential of Holder, Issuer, as well as trusted identity authentication database.

The micro-grid blockchain distributed identity authentication system can have multiple CA certification centers, usually by the power grid agency or other agencies. In addition, there are many terminal nodes in the microgrid.

16.4 BLOCKCHAIN-BASED DISTRIBUTED IDENTITY AUTHENTICATION SYSTEM

16.4.1 Symbol Definition

The definitions of the symbols in this section are shown in Table 16.1.

16.4.1.1 DID

DID can also be recorded as "did:", refer to Bitcoin's double hash way, the design of the DID identifier can use base58 (ripemd160 (sha256 (<Public Key>))) calculation method, where Public Key is the user's public key. Concrete form of DID identifier can be represented as "did:1:0x0086eb1f712ebc6f1c276e12ec21".

16.4.1.2 Subject

Subject represents the subject object and is the subject pair described by the claim elephant. Examples of entities include people, animals, and things.

TABLE 16.1
Symbol Definition

Identifier	Definition
DID	Distributed Identity Identifier [9]
DID Document	DID document, including DID public key, service endpoint, usage method, and other related attributes
Subject	Represents the subject object, which is the subject object described by the declaration. Examples of entities include people, animals, things, etc.
Claim	A statement, which is a statement about an entity. Declarations are expressed using a subject-attribute-value relationship.
Verifiable Credential	Electronic credentials that follow the W3C Verifiable Credential specification to contain one or more claim (Beuchat et al., 2010).
Presentation	Verifiable display of one or more verifiable credentials from the same principal certified data.

16.4.1.3 Claim

Claim refers to the claim in the user certificate, which is held by oneself. There are declarations of properties, content, etc.

16.4.1.4 Verifiable Credential

Verifiable credential refers to the application and issued credentials, consisting of metadata, claims, and attestations.

16.4.1.5 Presentation

Verifiable Credential displays one or more verifiable certificate data (Bhambri et al., 2022).

16.4.2 Algorithm Description

Trusted identity authentication systems include digital identity management systems and credential management system. In which, the digital identity management system includes identity registration stages, authentication stages, KYC process; and the credential management system includes Credential Creation phase, Credential Creation phase, Credential Verification phase segment, Voucher Update stage, Voucher Revocation stage.

16.4.2.1 Digital Identity Management System

A – Identity Registration Stage

$$createDid\ (addr) \rightarrow \{pk, sk, did\}:$$

Registration is the process of obtaining an identity, which needs to be done by the machine on the chain establishment to recognize the identity of the owner, performed by the identity owner.

Enter the address addr of the user node, and the system creates the public and private keys (pk, sk) and the registered user identity identifier did on the chain and return (did the public key address can be used directly). The public and private keys of the system user Denote as (pk, sk) u u, and denote the public and private keys of the issuer as (pk, sk) c c, The validator's public and private keys are (pk, sk) v v.

B – Authentication Phase

$$verifyDid\ (did,\ sign,\ pku) \rightarrow result$$

result: Verification is to verify whether the provided identity is legitimate the process of serving. The identity owner provides the identity identifier did, related the signature sign and identity public key pku of the challenge message, sk $sign = sign\ (ch)$ u, ch is the challenge for the validator. Validator in Query the existence of did on the chain, and use the public key pku to verify the user's signature. The correctness of the name sign, and then call the smart contract to query DID attribute, and returns the query result, where result={TRUE/FALSE}. TRUE means authentication passed, and FALSE means authentication failed.

C – KYC Process

The alliance committee provides users with the identity information verified, and after the verification is passed, it is encrypted and stored in the data. In the database, it is used for subsequent supervision and user credential information comparison and verification to prevent illegal transactions.

16.4.2.2 Credential Management System

A – Credential Creation Stage Create

Credential $(did, a, v) \rightarrow cred$: The credential creation algorithm is executed by the issuer of the credential. The DID holder provides its identity identifier DID user attribute, and the corresponding attribute value v. A after verifying the attribute, the issuer first generates $claim_i\ \{a, v\}$. Then, compute the attribute value summary, and generate a signature proof sk =Sign ( ) c Sh for the claim signature, h = H (v), where H is the hash function; then, the issuer.

Add credential authorization time and credential update time to authorization certificate time, user-related attributes, how to use user's DID, test certificate type (key type), port used, update properties and metadata information such as the issuer's signature information to complete authorization.

In case of creation and issuance of certificates, if the attribute held by the user is a privacy attribute, select generator is $g \in G$ and computes $A = g^a\ (mod\ q)$. Blindness Child r, its commitment value $c = g^r$, so the anonymous declaration is $claim = \{a, A\}$ and sent to the sender. The issuer verifies the commitment value and the verification is successful then accept the promise; generate a signature proof S for the statement signature, sk =Sign ( ) c S c, and add the certificate authorization time, Software-related attributes, the user's own attributes, and the issuer's signature information to complete the creation and issuance of the authorization certificate.

Pick (ch, rp) As a result, the proof can be verified by other users and validators certificate. Common Credential Creation Algorithms and Private Credential Creation Algorithms as shown in Algorithm 1 and Algorithm 2.

ALGORITHM 1 COMMON CREDENTIAL CREATION ALGORITHM

Enter the user identity identifier DID attribute a, and the corresponding attribute

Sex value v

output document result

check(did) //Check the existence of did

for a_i in a, v_i in v: //Load user attributes

setClaim (a_i, v_i) //Set the claim value

$h_i = H(v_i)$ // Calculate attribute value

End for

$h = h_1 || \ldots\ldots || h_n$

Connection digest value

$S = Sign_{sk_e}(h)$ /Calculate the signature value

end

ALGORITHM 2 PRIVACY CREDENTIAL CREATION ALGORITHM

Enter the user identity identifier DID attribute a, and the corresponding attribute

value v

output document result

check(did) //Check the existence of did

select $g \in G$ //select group parameters

$A = g^a \pmod q$ //compair the commitmeny of a

setClaim(a, A) //Set the claim value

select $r \in Z_q$ //select blind factor r

$c_i = g^r \pmod q$ //countCalculate r of Inherit promise

ch = H(g, A, t) //Calculation challenge

$rp = v - ch \cdot a \pmod q$ //calculate the response

h = H(A) //summary of computed property values

$sk = Sign_{sk_e}(h)$ //Calculate the signature value

cred = {claim, h, S} //Create credential

end

Evidence creation stage create $Evidence(object, sk_c) \to ev$. Enter the object that is not on the chain (usually all issued credential objects) and identity private key skc, the output of the algorithm is stored in evidence ev.

Calculate the hash value of the incoming object, through the body share the private key skc signature $s_0 = Sign_{sk_e}(h_0)$, generate the deposit evidence $ev = \{h_0, s_0, ex\}$ is other auxiliary message or remark message and put the deposit evidence ev on the chain. The depository creation algorithm can be created by the holder or issuer calls.

B – Credential Verification Phase

verifycredential (*did*, *cred*, *pak*) result: Enter the holder's identity did, one or multiple credentials is cred, and identity public key pk to be verified, the verifier first Reconstruct credential digest from credential $h_{0's_0}' = H(cred)$, then use the issuer's public key pk_c verifies the certificate's signature s_0 and matches the signature s_0 is compared with the digest h_0 to compare whether h_0' is consistent with h_0 Wait. Then return the verification result, where *result* = {*TRUE* / *FALSE*} TRUE indicates that the credential verification is successful.

However, FALSE indicates that the credential validation failed. The method is shown in Algorithm 3.

ALGORITHM 3 COMMON CREDENTIAL AUTHENTICATION ALGORITHM

Input User ID, did, cred, public key pk

output TRUE / FALSE

check(did) // check the existence of did

$tag_1 = verSign_{sk_e}(s_0)$ //Verify the signature value

$h_1' = H(cred)$ //reconstruct credential digest

$tag_2 = isEqual(h_0, h_0')$ //Read the digest h_0 from the chain, and compare with h_0'

return $tag_1 \&\& tag_2$

end for

end

In the zero-knowledge credential verification algorithm, in addition to the regular input some zero-knowledge auxiliary parameters need to be entered to support the zero-knowledge verification of exposed attribute values, in addition to the regular credential verification steps.

Steps are also required to verify the zero-knowledge challenge and response, and from the chain. Read the digest and compare it with the digest data recovered by the algorithm. The specific verification is shown in Algorithm 4.

ALGORITHM 4 ZERO-KNOWLEDGE CREDENTIAL VERIFICATION ALGORITHM

Input User ID, did, cred, public key

$pk, challenge/response\ value\ (ch, rp), commitment\ A$

output TRUE / FALSE

$check(did)$ // check the existence of did

$tag_1 = verSign_{sk_e}(s_0)$ //Verify the signature value

$h_1' = H(cred)$ //reconstruct credential digest

$tag_2 = isEqual\ (h_0, h_0')$ //Read the digest o h from the chain.

and compare with h_0'

$t' = g^r A^{ch}$ // refactoring promise

$tag_3 \leftarrow ch\ ? = H(g, A, t')$ //Compare equation

$return\ tag_1\&\&tag_2\&\&tag_3$

end for

end

End-for-end credential verification consists of two modes, namely off-chain mode and on-chain mode. In off-chain mode, the verifier only needs to verify the signature of the credential name, validity period, and whether to revoke and other attribute items to judge the authorization certificate authenticity, which is suitable for scenarios with low security requirements or offline/emergency scenarios; in the on-chain mode, in addition to the above verification, the user also needs to reverse the construction of the deposit certificate and call the smart contract to check on the chain query and compare the certificates to prove consistency with the certificates held on-chain. This mode is suitable for scenarios with high security requirements.

C – Credential Update Phase

$updateCredential(did, cred, newClaim) \rightarrow newCred$: Enter the holder's identity did, the credential cred and the new claim newClaim, $newClaim_i = \{att_i', val_i'\}$ The issuer verifies the original certificate cred and the new attribute declares new claim after updating the authorization certificate's update attribute, version number and declaration body, and adds a signature letter message to complete the update of the authorization certificate credential revocation phase.

credential Subject cred => credential body cred: Enter the holder's identity DID and credentials Subject cred, after the issuer verifies the original certificate cred, updates the authorization. The revoke attribute of the certificate and re-adding the signature information will revoke the sales certificate number cid and the deposit certificate are on the chain. Then return the result of the cancellation result, where *result* = {*TRUE* / *FALSE*}. TRUE means the credential revocation succeeded, and FALSE indicates that the credential revocation failed.

16.4.3 System Application

The blockchain-based trusted distributed identity authentication system for power grids is applied to the distributed transaction system of the micro-grid. By using DID, the user holds a globally unique identifier, and DID identifier symbol can be created the user themselves, manage their own information. So, users can fully own, control, and manage their identities.

Since the user owns the identity information, the user and the verifier lack of trust between apps. Therefore through blockchain technology, it is possible to confirm the information published by each node user and to establish the trust between the nodes of the information on the chain.

Thanks to the distributed ledger technology of the blockchain, users do not need to in every grid trading system (even off-grid systems). When registering an account, as long as the user and the verifier are in the same chain.

A comprehensive identity can be obtained through distributed user social relations authentication; verifiers can perform identity to grid users through attributes access, and user must have the corresponding attributes to access.

Although the user's digital identifier cannot infer the user's true identity information, the credentials may contain the user's privacy information.

For sensitive identity information, users can configure it as a discrete logarithmic model and encrypted as a commitment to hide the User privacy, the verifier only needs to pass the zero-knowledge proof technology, and effective verification to achieve privacy protection for expired or in question credentials, users and administrators can revoke password credentials.

In addition, the power grid trusted distributed identity authentication system based on blockchain can realize cross-platform porting, such as compatibility with systems outside the grid system authentication (Chumachenko et al., 2022).

16.5 EXPERIMENTAL EVALUATION

This chapter deploys FISCO on a CentOS host BCOS and identity authentication system server, laboratory host model. It has been verified that the distributed access of the power grid is simulated by simulating the power grid client process to verify the efficiency and availability of the system in this chapter. The parameters of the host computer are: Intel(R) Core i7-7700 CPU @ 3.60 GHz 3.60 GHz, RAM 16.00 GB, CentOS 7 system. The parameters of the client host is Intel(R) Core i5-4590 CPU @ 3.30 GHz 3.30 GHz, RAM 8.00 GB. Using the Apache JMeter test tool to test the system separately each interface of the Http request test is performed, and the number of all experiments as shown in Figure 16.2. Diagram shown the average of 10 runs and the number of nodes deployed by the blockchain for 6 nodes.

16.6 SMART CONTRACT DESIGN

The smart contracts in the system are divided into DID contracts from data objects (Khang et al., Data-Driven Blockchain Ecosystem, 2022). They are Smart Contract, Credential contract, CPT (credential template contract); According to the definition

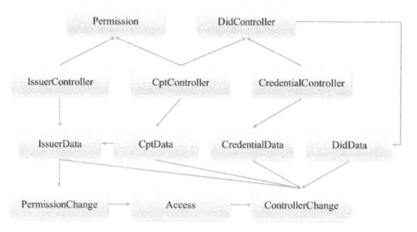

FIGURE 16.2 Contract dependencies.

and design dimension of the structure, it can be divided into Did Data contract, Credential Data contract, Cpt Data contract, Issuer Data Contracts, and other roles and permission control contracts for objects, which depend on the dependency relationship as shown in Figure 16.3.

The DID smart contract is responsible for the establishment of the ID system on the chain, including generating DIDs, generating DID-related documents, read DIDs on the chain and fetching and updating. The authority smart contract is responsible for consortium chain authority management, including the definition of DID roles on the chain, and the determination of operations and permissions.

16.7 SYSTEM PERFORMANCE TEST

During the system performance test, the simulated client first initiates the identity a registration request, that is a DID creation request and the number of concurrent users as Table 16.2.

The time trend of DID creation is shown in Figure 16.3. It can be seen that the identity creation time increases with the number of concurrent users.

The change of the statement creation time with the number of user attributes is shown in Figure 16.4.

It can be seen that the statement creation time varies with the number of user attributes, basically showing a linear trend, but the increasing trend is relatively weak. In fact, the number of attributes in the distributed credential declaration is limited, no more than 20, so the system fully meets the actual need to run as shown in Table 16.3.

The creation time of common credentials under different number of concurrent users and the time of zero-knowledge voucher creation are shown in Figure 16.5 and Figure 16.6, respectively.

TABLE 16.2
Time Trend of DID Creation

Concurrent Users	Time in ms
20	1475
40	1600
60	1700
80	1750
100	1752

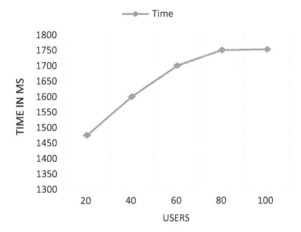

FIGURE 16.3 Time trend of DID creation under different number of concurrent users.

TABLE 16.3
Creation Time Data

Users	Time in ms
20	15
40	20
60	19
80	21
100	23

As can be seen from Figure 16.5 and Figure 16.6, the voucher creation time increases with the increase in the number of concurrent users, but the growth rate is gradually weakened. And the performance of zero-knowledge credentials is almost the same as that of ordinary credentials and the performance can meet the actual needs.

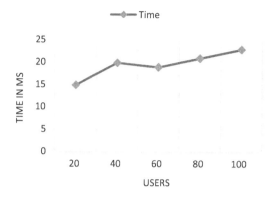

FIGURE 16.4 Changes in statement creation time with the number of user attributes.

TABLE 16.4
Data for Creation Time of Common Credentials

Users	Time in ms
20	20
40	35
60	55
80	85
100	105

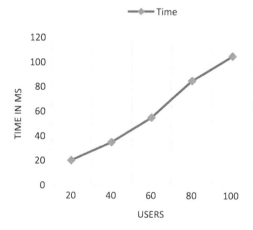

FIGURE 16.5 The creation time of common credentials under different number of concurrent users.

Blockchain-based Privacy Protection Credential Model

Although users, the number of concurrency is increasing, but the distributed credential issuance method is decreasing, lower overall system overhead, compared to centralized credential issuance way, the efficiency has been improved. In the actual system, the credentials creation is undertaken by different committee nodes, which are actually run submodules independently and in parallel without interfering with each other, without getting tired when included in the total system time as Table 16.4.

In Figure 16.6, the time of zero-knowledge credential creation under different number of concurrent user's different concurrent users is shown. The time and zero of common credential verification under different number of concurrent users is the time of knowledge certificate verification is shown in Figure 16.6. The time of knowledge certificate verification is shown in Figure 16.6. Similar to credential creation, the credential verification time varies with the number of concurrent users but the growth rate gradually weakens.

Because the user, the number of concurrency is increasing, the distributed credential verification method is decreasing. The overall overhead of the system is more

TABLE 16.5
Creation Verification Time of Common Credentials

Users	Time
20	25
40	185
60	400
80	585
100	710

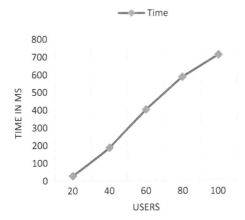

FIGURE 16.6 Time of common credential verification under different number of concurrent users.

efficient than in the centralized verification method. The overall situation is within the acceptable range as seen in Table 16.5.

The distributed certificate generation and revocation method can reduce the total system cost. Compared with the centralized verification method, the efficiency is improved. As can be seen from Figure 16.6, the credential update time varies with user attributes increases with increasing number, in a linear trend, due to each attribute, the magnitude limit of the increasing trend is weak. In practical applications, the number of sexes is limited and usually single digits of the time is completely full meet the needs of actual operation.

16.8 CONCLUSION

In order to address the trusted access of terminal nodes or applications in the microgrid and business access control, this chapter used FISCO BCOS blockchain and a distributed identity management that supports multi-center platform and autonomous control of user identity and multi-platform, cross-chain, migration of identities and credentials across applications.

The user fully owns, controls, and manages their own identity for microgrid systems and upper-layer applications. Build an open, transparent and credible distributed identity bottom frame, the framework can apply identity autonomy and fine-grained access control. In the future, we will focus on the privacy protection of the authentication process for any system.

REFERENCES

Abbas, Gardashova Latafat, Khang, Alex, Vladimir Hahanov, & Vugar Abdullayev Hajimahmud. Cyber-physical-social system and incident management, *AI-Centric Smart City Ecosystem: Technologies, Design and Implementation*, 25–44, 2022, doi: 10.1201/9781003252542-2

Beuchat, J.L., J.E. Gonzalez-Diaz, S. Mitsunari, E. Okamoto, F. Rodriguez-Henriquez, & T. Teruya. High-speed software implementation of the optimal ate pairing over Barreto-Naehrig curves. *Proc. 4th Int. Conf. Pairing-Based Cryptography*. 2010 LNCS 6487, pp. 21–39.

Bhambri, Pankaj, Khang, Alex, Sita Rani, & Aman Kataria. *Big Data, Cloud Computing and Internet of Things*, ISBN: 978-1-032-284200, 2022, doi: 10.1201/9781032284200

Bhambri, Khang, A., Sita Rani, & Gaurav Gupta, *Cloud and Fog Computing Platforms for Internet of Things*, ISBN: 978-1-032-101507, 2022, https://doi.org/ 10.1201/9781032101507

Boneh, D., & M.K. Franklin. Identity-based encryption from the Weil pairing, *Proceedings of the Advances in Cryptology (EUROCRYPT'01)* vol. 2139, pp. 213–29. 2001.

Chen, L.Q. Constructing a safe, reliable, economical and clean modern energy industry system [J]. *Journal of Wuhan University of Science and Technology (Social Science Edition)*, 2011, 13(5), 497–505.

Chumachenko, Svetlana, Abdullayev Vugar Hajimahmud, Khang, Alex, Vladimir Hahanov, Eugenia Litvinova, & Abuzarova Vusala Alyar. Autonomous Robots for Smart City: Closer to Augmented Humanity. *AI-Centric Smart City Ecosystem: Technologies, Design and Implementation*, 161–76, 2022, doi: 10.1201/9781003252542-7

Hahanov, Vladimir, Eugenia Litvinova, Khang, A., Svetlana Chumachenko, Vugar Abdullayev Hajimahmud, & Abuzarova Vusala Alyar. The Key assistant of smart city – sensors and tools. *AI-Centric Smart City Ecosystem: Technologies, Design and Implementation*, 361–73, 2022, doi: 10.1201/9781003252542-17

Hajimahmud, Vugar Abdullayev, Khang, Alex, Nazila Ali Ragimova, & Abuzarova Vusala Alyar. Advanced technologies and data management in the smart healthcare system. *AI-Centric Smart City Ecosystem: Technologies, Design and Implementation*, 348–60, 2022, doi: 10.1201/9781003252542-16

Hong, Z. The strategic position of electric power in China's energy [J]. *China Economic & Trade Herald*, 2014 (19), 10–16.

Kataria, Aman, Alex Khang, Sita Rani, & Pankaj Bhambri. Smart city ecosystem: Concept, sustainability, design principles and technologies. *AI-Centric Smart City Ecosystem: Technologies, Design and Implementation*, 1–34, 2022, doi: 10.1201/9781003252542-1

Khan, S., A. Jadhav, I. Bharadwaj, M. Rooj, & S. Shiravale. Blockchain and the identity based encryption scheme for high data security. *2020 Fourth International Conference on Computing Methodologies and Communication (ICCMC)*, 2020, 1005–8, doi: 10.1109/ICCMC48092.2020.ICCMC-000187

Khang, A., Geeta Rana, Ravindra Sharma, Alok Kumar Goel, & Ashok Kumar Dubey, The role of artificial intelligence in blockchain applications, *Reinventing Manufacturing and Business Processes through Artificial Intelligence*, 19–38, 2021, https://doi.org/10.1201/9781003145011-2

Khang, A., Pankaj Bhambri, Sita Rani, & Gaurav Gupta. *Cloud and Fog Computing Platforms for Internet of Things*. Chapman & Hall. ISBN: 978-1-032-101507, 2022, doi: 10.1201/9781032101507

Khang, A., Chowdhury, S., & Sharma, S. (Eds.). (2022). *The Data-Driven Blockchain Ecosystem: Fundamentals, Applications, and Emerging Technologies* (1st ed.). CRC Press. https://doi.org/10.1201/9781003269281

Khang, A., Rani, S., & Sivaraman, A.K. (Eds.). (2022). *AI-Centric Smart City Ecosystems: Technologies, Design and Implementation* (1st ed.). CRC Press. https://doi.org/10.1201/9781003252542

Niux J., & Lic, H. On China's strategic orientation, policy framework and role of government in renewable energy sources [J]. *Chinese Public Administration*, 2014 (3): 100–3.

Shobana, G., & M. Suguna. Block chain technology towards identity management in health care application. *2019 Third International conference on I-SMAC (IoT in Social, Mobile, Analytics and Cloud) (I-SMAC)*, 2019, 531–5, doi: 10.1109/ISMAC47947.2019.9032472

Shobanadevi, A., Tharewal, S., Soni, M., et al. Novel identity management system using smart blockchain technology. *Int J Syst Assur Eng Manag* (2021). https://doi.org/10.1007/s13198-021-01494-0

Sita Rani, Khang, Alex, Meetali Chauhan, & Aman Kataria, IoT equipped intelligent distributed framework for smart healthcare systems, *Networking and Internet Architecture*, 2021, https://arxiv.org/abs/2110.04997v2, https://doi.org/10.48550/arXiv.2110.04997

Soni, M., & D.K. Singh. Blockchain implementation for privacy preserving and securing the healthcare data. *2021 10th IEEE International Conference on Communication Systems and Network Technologies (CSNT)*, 2021, 729–34, doi: 10.1109/CSNT51715.2021.9509722

Tu, Y., J. Gan, Y. Hu, R. Jin, Z. Yang, & M. Liu. Decentralized identity authentication and key management scheme. *2019 IEEE 3rd Conference on Energy Internet and Energy System Integration (EI2)*, 2019, 2697–2702, doi: 10.1109/EI247390.2019.9062013

Wang, W, Lir, & Jiang, J.C. Key issues and research prospects of distribution system planning orienting to energy internet [J]. *High Voltage Engineering*, 2016, 42(7), 2028–36.

Wu, Z., Y. Xiao, E. Zhou, Q. Pei, & Q. Wang. A solution to data accessibility across heterogeneous blockchains. *2020 IEEE 26th International Conference on Parallel and Distributed Systems (ICPADS)*, 2020, 414–21, doi: 10.1109/ICPADS51040.2020.00062

Yan, X., Y. Lu, C.-N. Yang, X. Zhang, & S. Wang. A common method of share authentication in image secret sharing. *IEEE Transactions on Circuits and Systems for Video Technology* 31(7), 2896–2908, July 2021, doi: 10.1109/TCSVT.2020.3025527

Yong, Yuan, & Wang Feiyue. Blockchain: The state of the art and future trends. *Acta Automatica Sinica* 42(4), 481–94, April 2016.

Zhaoy, H., Peng, K., Xub, Y., et al. Status and prospect of pilot project of energy blockchain [J]. *Automation of Electric Power Systems*, 2019, 43(7), 14–22, 58.

Zhouh, Y., Qianw, H., Baij, J., et al. Typical application scenarios and project review of energy blockchain[J]. *Electric Power Construction*, 2020, 41(2), 11–20.

Zou, B., G. Zhao, H. Tang, R. Nie, R. Huang, & J. Tang. Archives chain: Distributed PKI archives system. *2021 4th International Conference on Advanced Electronic Materials, Computers and Software Engineering (AEMCSE)*, 2021, 1009–13, doi: 10.1109/AEMCSE51986.2021.00206

Index

A

agriculture sector 197, 198, 201
artificial intelligence 32, 34, 119, 120, 189–91, 197, 199, 200–2
authentication 114, 215–17, 219, 220, 249–51, 253, 255, 259, 260
authorization 52, 66, 144, 215, 256, 258, 259
autonomous control 249, 251, 264

B

banking 32–4, 43, 84, 85, 87, 90, 94, 97, 115, 119, 170
Be Credible, Open & Secure (BCOS) 249, 251, 260, 264
big data 22, 32, 46, 62, 124, 138, 192, 197, 227
billing system 156, 163, 167
bitcoin 84, 90–2, 98, 103, 113–15, 117–18, 123, 208, 209, 211, 214, 216, 219–21
blockchain-based cloud 235, 240, 242, 243
blockchain framework 62, 93
blockchain platform 14, 24
blockchain process 45, 56
blockchain security 216, 220–2
blockchain system 2, 4, 5, 12, 14–16, 172, 174, 179, 122, 124
blocks 128, 130, 208, 209, 211, 214, 216, 219, 222

C

centralized system 250
citizen management 34, 40, 48
cloud computing 32, 46
cloud platform 229, 231, 240, 242, 243
cloud platform operators (CPO) 231, 232, 235, 236, 239
computer technology 108
consensus mechanism 130, 211
consensus problem 209
consensus issue 209
consortium chain 5, 230, 261
consortium blockchain 5, 63, 64, 67, 78
credential verification 255, 257, 258, 262, 263
cross-chain 252, 264
cross-department 251
cross-organization 251
cross-shared transactions 170
cryptocurrency 21, 23–9, 207, 209, 213, 216, 221, 222; industry 52

cryptography 3, 13, 53, 118, 120, 123, 157, 165, 170, 209, 212, 213, 215, 220; algorithms 74; storage 26

D

database 190, 195, 208, 209, 212, 222
data network 190
data of things (DoT) 34
data privacy 84, 86, 104, 130, 217, 218
data quality 105
DDoS attacks 54
decentralized data 15, 208, 209
decentralized framework 242
decentralized identifier (DID), 251, 253–257, 259, 269, 261
decentralized network 67, 87, 230
decentralized technology 190
deep learning 42, 44, 48
digital currency 90, 92, 94
digital economy 94
digital identity 95, 118, 201, 250, 251, 255
digital ledger 84, 87, 156
digital signature algorithm 229, 232, 242
disease detection 192, 199
disease prediction 196, 199, 201
distributed database 4, 5, 10, 107
distributed ledger 22, 24, 84, 86, 93, 209, 222, 253, 259
distributed network 209
distributed storage 170, 208
distributed system 248
distribution system 194, 198, 201

E

E-commerce 26, 86, 98
E-government 93, 213
electronic data 53, 116
encryption algorithm 157, 160, 165
encryption system 3, 13
encryption technology 107
energy 248, 249
ethereum 10, 11, 13, 24, 52, 55, 56, 87–9, 93, 130, 159, 211, 219, 220, 230, 239, 250
extra currency 53

F

finance 3, 34, 41, 44, 89, 94, 96, 97, 108, 120, 138, 140, 170

Index

financial industry 207
financial system 51, 84, 90, 92
Financial Blockchain Shenzhen Consortium (FISCO) 249, 251, 260, 264
framework 215, 220, 250, 264
fraud detection system (FDS) 36

G

genesis block 73, 74
Genetically Modified Firefly Optimization Algorithm (GMFOA) 163–7
global positioning system (GPS) 32, 34, 37, 42

H

hacker 2, 12, 27, 52, 258
hashed/hash-based message authentication code (HMAC) 162, 163
health care 190, 191, 197, 202
health information 127, 128, 130
hybrid Blockchain 5
hyperledger 220, 230; fabric 67, 69, 76, 79, 128, 131, 132, 159, 220

I

information and communications technology (ICT) 34, 108, 109
intelligent applications 207, 208
internet of things 33, 190, 191, 197, 138, 140, 202, 213, 215, 227, 235, 248
internet security 28
IoT light OS driven by blockchain technology (IoTChain) 156, 159, 167, 215
IoT platform 190
IoT technology 33, 201, 222

K

key performance indicator (PKI) 250, 252
know your customer (KYC) 96, 249, 251, 253, 255

L

life insurance 26
litecoin 24, 51, 87, 88
logistics 41, 138, 170, 229

M

machine learning 42, 43, 124
marketing strategy 201
MATLAB 230, 240, 241
medical data 95, 127, 130, 138, 140, 144, 148, 215
microgrid 248–50, 253, 264
mining pool 220, 221

N

node random 180
non-fungible token (NTF) 91, 92

O

omniledger 181–5
optimization 117, 130, 141, 144, 146, 157, 161, 213, 228, 243

P

P2P network 13, 170, 209, 222
peer-to-peer 4–6, 15, 56, 191, 208, 209, 217, 222
personal privacy 249
physical hardware 26
Practical Byzantine Fault Tolerance (PBFT) consensus algorithm 172, 185
private data 16, 124
private key 6, 9, 11, 14, 22, 26, 53, 55, 75, 118, 120, 180, 212, 217, 219, 230, 233, 253, 255, 257
privacy protection 249, 259, 264
privacy and security 130, 190, 208, 217
private blockchain 5, 6, 67, 79
private chain 220, 230
private-key 212
probability analysis 171, 173
proof of work (PoW) 52, 214, 218, 220, 221
public blockchain 5, 6, 67, 79
public chain 229, 230
public key 53–5
public ledger 53, 208, 209, 219

R

radio frequency identification (RFID) 34, 156–9, 162
real-time data 11, 32, 87, 124

S

safe work method statements (SWMS) 36
secure hash algorithms (SHA) 8, 14
security 156, 158, 165, 248–50, 253, 259, 264
security assertion markup language (SAML) 250
self-learning 195
single sign-on (SSO) 250
smart agriculture 32, 33, 42
smart banking 32, 33, 43
smart buildings 32, 33, 38
smart contract 4–7, 17, 22, 62, 69, 71, 72, 75–9, 103–6, 131–3, 157, 159, 209, 211, 212, 230, 231, 235 7, 239, 242, 250, 255, 259–61

Index

smart education 33, 34, 36
smart energy 32, 33, 41, 47
smart environment 32, 33, 36
smart healthcare 138, 140, 141, 151
smart shopping 156–8, 163
smart technology 32
smart transport 32, 34, 48
software development 212
spyware 28
supply chain 85, 93, 94, 190, 191, 201, 202
symmetric keys 159
synchronous discrete twofish encryption algorithm (SDTEA) 163–7

T

terminal nodes 248, 249, 253, 264
total workforce management services (TWMS) 36
traditional currency 84, 87
transaction data 216, 217
transaction per second (TPS), 171, 176, 179, 181, 183–6
transaction process 74–7
transaction rollbacks 171, 174, 176
trusted access 249, 250
trustworthy 24, 78, 88, 104, 107, 121, 124

U

unified identity 252, 253
unspent transaction outputs, unspent transaction output (UTXO) 171–4

V

vehicle service 35, 36
verifiable random function, verifiable random function (VRF) 180, 181
virtual currency 23, 27, 53
virtual money 53, 92
vulnerability risk, 220

W

wallets 23–8, 212
web application 194
webcam 47
web service management system (WSMS) 36
Wi-Fi 39
wireless sensor 32
workflow 69, 88, 117, 121, 196, 229

Z

zero-knowledge 181, 249, 258, 259, 262

Taylor & Francis eBooks

www.taylorfrancis.com

A single destination for eBooks from Taylor & Francis with increased functionality and an improved user experience to meet the needs of our customers.

90,000+ eBooks of award-winning academic content in Humanities, Social Science, Science, Technology, Engineering, and Medical written by a global network of editors and authors.

TAYLOR & FRANCIS EBOOKS OFFERS:

- A streamlined experience for our library customers
- A single point of discovery for all of our eBook content
- Improved search and discovery of content at both book and chapter level

REQUEST A FREE TRIAL
support@taylorfrancis.com